山乘自然

心象自然

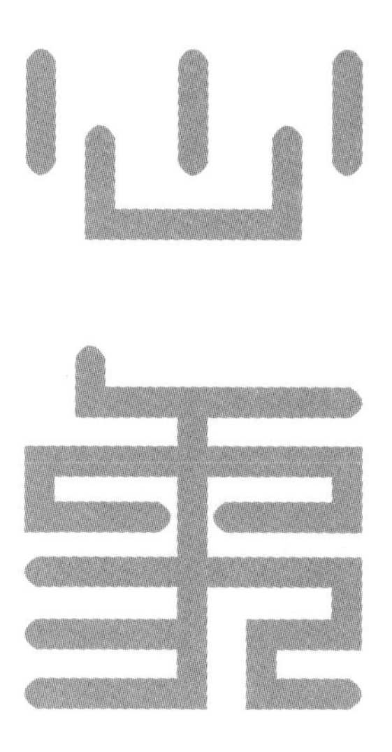

Inner Nature Imagery

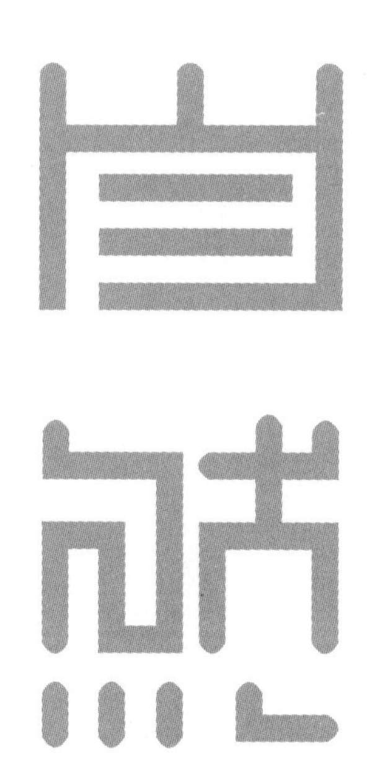

李存东　史丽秀　著

中国建筑工业出版社

心象篇

心象自然篇

Inner Nature
Imagery

Inner
Imagery

序一

意与形

《心象自然》这本书是以李存东、史丽秀领衔的中国建筑设计研究院园林景观团队多年耕耘的成果。书中不乏大量的设计实践，尽显园林景观之美，更可贵的是，他们把这些创作案例按照“心象自然”的理念加以归类整理，形成带有方法论意义的理论探索，旨在反映他们在园林景观设计中的心路历程。书中结合案例，对主观与客观、人与自然二元关系进行的深度思考，尤其从园林景观专业角度看，很有针对性，试图以图文结合的方式，表达自己的设计观念。

本书作者在大量园林景观设计实践基础上，选取了一些有代表性的典型案例，类型十分丰富，涵盖了大型公共建筑和民用居住建筑、城市广场和风情商街、新区开发和老城更新、会展中心和交通枢纽、科技园区和商业中心、大学校园和科研院所、园艺博览和遗址公园……通过对这些大量园林景观设计项目的整理归纳，寻找在不同类型案例中寓藏着的共同规律，使其上升为具有理性意义的观念和理论，反映出作者在园林景观创作中的理念和专业价值取向。

作者开门见山，在介绍案例之前，首先提出了“作品后面的设计观”，文字虽然笔墨不多，仅一页之幅，但言简意赅，一如本书书名——“心象自然”，这正是作者想要表达的主题和思想。我们知道，无论是规划、建筑，还是园林景观，任何一种设计，都离不开设计人的创意和相对物所处的环境。创意带有主观性，正所谓意在笔先，内化于心，关键字是“意”；环境则是客观存在，设计必须尊重环境、适应环境，进而改善环境、提升环境，设计的表达方式也将外示于景，关键字是“形”。创意即“心象”，环境乃“自然”。由此可见，规划和设计其实质就是力求处理好“心象”与“自然”的关系，把“天人合一”“道法自然”的传统理念，在当下的设计实践中得以传承和创新，正如作者所说，“‘心象自然’就是要平衡人与自然的关系”“既要延续自然界的生态准则，又要保持人类文明持续健康发展”，这应该是广大设计者应秉持的共同观念与态度、方法与追求。

我们知道，理论来源于实践，任何学科的理论都是在实践中得到的对事物的本质和规律的认识，而且这是一个深化认识和创新实践不断相互反馈的过程，因为事物总是在向前发展，赋予了实践以新的内涵和要求，其观念和理论必将进行修正和完善，唯此才能得到新的认识和正确的理论，才会有科学的决策和方法，从而更好地推动和指导实践。我国已进入高质量发展的新时代，对人居环境建设提出了新的更高的要求，人民群众对美好生活有了更多的期待。党的十九大对建设生态文明和美丽中国，既有战略性的部署，又有具体的目标要求，按照中央绘制的蓝图，到 2035 年基本实现社会主义现代化，生态环境根本好转，美丽中国目标基本实现；到 21 世纪中叶（2050 年），建成富强民主文明和谐美丽的社会主义现代化强国。这里“美丽”已成为我们奋斗目标的必要条件，这意味着我们从事的城乡建设、规划设计、园林景观工作肩负着重要的使命和责任。

中国有着“天人合一”“道法自然”的历史文化传统，这在新时期有了更具时代特征的清晰表达，把“人与自然和谐共生”确立为建设中国特色社会主义的基本方略之一，美丽中国必须以绿色价值观为引领，践行“绿水青山就是金山银山”的重要理念。我们园林景观专业人员，更应该在专业实践中认真贯彻切实体现这些基本要求，这里的关键是要强调重视设计工作、增强精品意识、提高设计的质量和水平，要鼓励大胆创意，创作出既能传承文脉，又能体现时代精神的优秀作品，提升园林景观的文化品位和艺术含量。从本书介绍的这些案例可以看出，设计者的创作态度是认真的，并在设计方法创新、注重设计质量、力求推出精品上下了不少功夫，是值得肯定的，可资作为同行们参考与借鉴，裨益大焉。

是为序。

宋春华
原建设部副部长
中国建筑学会原理事长

Foreword One

Meaning and Form

This book, Inner Nature Imagery, is the long-cultivated result of the garden landscape team of China Architecture Design & Research Group led by Li Cundong and Shi Lixiu. There are many design practices in the book, which show the beauty of the garden landscape. What is more valuable is that they categorize these creation cases according to the concept of "inner nature imagery", forming a theoretical exploration with methodological significance, aiming at reflecting their mental process in garden landscape design.The book combines cases to make a deep reflection on the dual relationship between subjectivity and objectivity, man and nature, and in particular, from the professional perspective of the garden landscape, very targeted, trying to use the combination of text and images, to express their own design concept.

On the basis of a large number of garden landscape design practice, the author selected some representative typical cases. The types are very diverse, covering large public buildings and civilian residential buildings, urban squares and commercial streets with amorous feelings, development of new areas and renewal of old cities, convention and exhibition center and transportation junction, science and technology park and commercial center, university campus and scientific research institutions, horticultural expo and heritage park···Through the collation and induction of a large number of garden landscape design projects, the common rules hidden in different types of cases are sought, to make the concepts and theories rise to rational significance, reflecting the author's philosophy and professional value orientation in the creation of garden landscape.

Before introducing the cases, the author come straight to the point, and first propose the "design concept behind the work". The text is not very long, only one page, but concise and comprehensive, just like the title of the book — "inner nature imagery", which is exactly the theme and idea the author wants to express. We know that no matter planning, architecture or garden landscape, any kind of design is inseparable from the designer's creativity and the environment of the relative objects. Creativity is subjective. Like the old saying is that meaning has the priority before pens and digest in the heart. The key word is "meaning". The environment exists objectively. Design must respect the environment, adapt to the environment, so that improve the environment and promote the environment. The expression of the design will also be shown in the scene, the key word is "shape". Creativity is "inner imagery" and the environment is "natural". Thus it can be seen that the essence of planning and design is to try to deal well with the relationship between mind and nature, and to pass on and innovate the traditional concept of "unity of man and nature" and "Taoism follows nature" in the current design practice. As the author says, "the 'inner nature imagery' is to balance the relationship between man and nature." "We should not only continue the ecological norms of nature, but also maintain the sustained and healthy development of human civilization." This should be the common ideas and attitudes, methods and pursuits that general designer should hold.

As we know, theory comes from practice, and the theory of any discipline is the knowledge of the essence and law of things obtained in practice, and it is a process that deepening the knowledge and innovating practice constantly couple back on each other. Because things are always moving forward, practice is endowed with new connotations and requirements. Its concept and theory will certainly carry on the revision and the consummation, only then can obtain the new understanding and the correct theory, only then can have the scientific decision-making and the method, thus better promotes and guides the practice. China has entered a new era of high-quality development, which has put forward new and higher requirements for the construction of human settlements, and the people have more expectations for a better life.The 19th National Congress of the Communist Party of China (CPC) has both strategic plans and specific goals and requirements for building an ecological civilization and a beautiful China. In accordance with the blueprint drawn up by the central committee, we will basically realize socialist modernization, fundamentally improve the ecological environment, and basically achieve the goal of a beautiful China by 2035. By the middle of this century (2050), China will become a great modern socialist country that is prosperous, strong, democratic, culturally advanced, harmonious and beautiful. Here "beautiful" has become a necessary condition for our objective of the struggle, and this means that the urban and rural construction, planning and design, garden landscape work that we are engaged in shoulder an important mission and responsibility.

China has a historical and cultural tradition of "unity of man and nature" and "Taoism follows nature" , and this has a clear expression with the characteristics of the times in the new era.To establish "harmonious coexistence between man and nature" as one of the basic strategies for building socialism with Chinese characteristics, a beautiful China must be guided by green values, and the important concept that "clear waters and green mountains are golden mountains" should be practiced. Garden landscape professionals, we should earnestly implement and practically embody these basic requirements in the professional practice. The key here is to emphasize the importance of design work, enhance the awareness of fine works, improve the quality and level of design, to encourage bold creativity, to create excellent works that can not only inherit the context, but also reflect the spirit of the times, to enhance the cultural taste and artistic content of the garden landscape. From the cases presented in this book, it can be seen that the creative attitude of designers is serious, and paid a large mount of hard work in innovating the design methods, paying attention to the design quality, striving to launch the high-quality works.It is worth affirming, can be used as the consult and reference for peer, which is very helpful.

Song Chunhua
The Former Deputy Minister of Housing and Urban-Rural Development of the People's Republic of China
The Former Chairman of Architectural Society of China

序二

书如其人，笃思勤行

一起出差去上海的路上，李存东把《心象自然》的书稿交给我，希望我能做个序，我欣然接受了。其实我很少给书作序，但这本书稍有不同，因为我不仅欣赏心象自然的观点和书中每一个项目，更重要的是，我太熟悉文字和图片背后的这些人，太了解这些年他们走过的既艰辛又精彩的道路。

初识李存东和史丽秀以及他们的团队是 2006 年，那时候景观所创立没几年，建成的项目还不多，在业界还没有名气。他们经常找到我，探讨团队发展方向和路径，每次见到他们，我都会被满满的信心和乐观向上的状态所感染。同样，在他们的感召下，景观所的每一个人都朝气蓬勃、充满活力。在跟院内建筑师合作时，他们尽心尽力、不计得失，凭借良好的专业水准和职业精神逐渐赢得了很好的口碑，那时候我就很看好他们。

景观所后来的发展印证了我当初的判断，这些年下来，他们完成了大量有影响的项目。从早期跟崔愷院士合作的首都博物馆、拉萨火车站、苏州火车站，跟李兴钢大师合作的鸟巢，跟陈同滨所长合作的世界文化遗产项目汉长安城未央宫遗址公园、湖南老司城遗址公园，和刘燕辉老总合作的北川、玉树灾后重建项目等，到后来他们自我拓展市场，布达拉宫周边环境整治、西单商业区、牙买加中国园林、万科、融创等住区景观，近几年更是作品频出，大项目不断，北京 APEC 会议雁栖湖会议中心南广场、2018 中国南宁国际园林博览会、2019 北京世界园艺博览会等等。随着这些项目的建成，他们的设计水准逐渐被行业认可，多年的积淀使得每一个作品都精益求精，可圈可点。

难能可贵的是，他们不仅创业成功，作品出色，在学术追求上也笃定前行，坚持不懈。这些年来，他们和北京林业大学、中央美术学院、北京建筑大学等多个学校合作交流、培养人才。与《中国园林》《风景园林》《景观设计》等多个杂志合作开展活动、发表学术文章。与中国建筑学会、中国风景园林学会、中国公园协会等行业组织合作，推动行业发展。作为牵头单位，在中国建筑学会成立园林景观分会，在中国公园协会成立规划设计分会，通过这些平台的建立，一方面他们的努力得到了行业的认可，另一方面他们也能够为行业进步尽一份责任。

这本书的出版又将是一个里程碑，作品背后设计思想的总结提炼是一个队伍走向成熟的标志。理论探索是当下社会最为需要的，尤其是对被称为“世界园林之母”的中国园林的研究，既迫切又有难度。“心象自然”可以说是对中国园林传承与创新的一个很好的探索，这种探索不是凭空臆想，也不是对他人设计的分析归纳，而是通过自己大量实践得到的真实感受，这种边实践边研究的方法是值得学习的。

传承中华文化，打造中国设计，促进科技进步，引领行业发展，这是我们多年总结的作为学术团体和科技企业的发展目标，李存东和史丽秀以及他们的团队正是这个目标非常直接非常真实的践行者。他们多年深耕，执着探索，他们的所思所想都通过这本书里的一个个鲜活的项目反映出来，对我来说，书如其人，倍感亲切。祝愿他们在实践探索和学术研究的道路上继续前行，把心象自然的理念不断丰盈完善，为构建中国园林的现代理论笃思勤行。

修龙
中国建筑学会理事长
中国建设科技集团原董事长

Foreword Two

Like Author Like Book, Think Seriously and Do Diligently

On a business trip to Shanghai together, Li Cundong gave me the manuscript of "Inner Nature Imagery", hoping that I could make a foreword, which I readily accepted. Actually, I rarely make foreword for books, but this one is a little different, because I can't help but appreciate the idea of nature in mind and every project in the book, more importantly, I am too familiar with the people behind the words and pictures, too familiar with the hard and wonderful road they have traveled over the years.

I first met Li Cundong, Shi Lixiu and their team in 2006, the landscape institute was just founded a few years ago at that time. The completed projects were not many, and it was not well-known in the industry. They often came to me to discuss the development direction and path of the team. Every time I saw them, I was always infected by their state that is full of confidence and optimism. Similarly, under their inspiration, everyone in the landscape institute was full of vigor and vitality. When I cooperated with the architects in their institute, they made every effort, regardless of gains and losses, and gradually won a good reputation by virtue of their good professional standards and professional spirit. At that time, I was very optimistic about them.

The subsequent development of the landscape institute confirms my original judgment. Over the years, they have completed a large number of influential projects. From the early works cooperated with academician Cui Kai such as capital museum, Lhasa railway station, Suzhou railway station, the Bird's Nest cooperated with master Li Xinggang, the works cooperated with president Chen Tongbin such as the world heritage project- Han Chang 'an Weiyang Palace Ruins Park, Hunan Laosicheng Ruins Park, to the works cooperated with chief designer Liu Yanhui such as Beichuan, post-disaster reconstruction projects in Yushu, then later they expanded their market on their own, such as surrounding environment renovation of the Potala Palace, Xidan business district, Chinese garden in Jamaica, and the residential landscape of Vanke, Sunac and so on. They created more works in recent years, and a lot of large projects, such as the south square of Yanxi Lake Conference Center for APEC meeting in Beijing, 2018 China Nanning International Garden Expo, 2019 Beijing International Horticultural Expo and so on. With the completion of these projects, their design standards are gradually recognized by the industry, and years of accumulation makes every work strive for excellence, worthy of praise.

What is most valuable is that they are not only successful entrepreneurs with excellent works, academic pursuit is also determined to move forward, with constant perseverance. Over the years, they have cooperated with Beijing Forestry University, Central Academy of Fine Arts, Beijing University of Civil Engineering and Architecture and other schools to exchange and train talents, have cooperated with Chinese Landscape Architecture, Landscape Architecture, Landscape Design and other magazines to carry out activities and publish academic articles, have cooperated with Architecture Society of China, Chinese Society of Landscape Architecture, Chinese Association of Parks and other industry organizations to promote the development of the industry. As the leading company, they have set up a branch of garden landscape in Architecture Society of China and a branch of planning and design in Chinese Association of Parks. Through the establishment of these platforms, on the one hand, their efforts have been recognized by the industry, on the other hand, they can also contribute to the progress of the industry.

The publication of this book will be another milestone, and the summary and extraction of the design ideas behind the work is a sign of the team's maturity. Theoretical exploration is the most needed in the current society, especially the research on Chinese garden which is known as the "mother of the world garden", is both urgent and difficult. Inner nature imagery can be said to be a good exploration of the inheritance and innovation of Chinese gardens. This exploration is not a figment or an analysis and induction of other people's designs, but a real feeling gained through a large number of their own practices. This method of studying while practicing is worth learning.

Inheriting Chinese culture, creating Chinese design, promoting scientific and technological progress, and leading the development of the industry are our development goals summarized over the years as an academic group and a technology enterprise. Li Cundong, Shi Lixiu and their team are very direct and real practitioners of this goal. Over the years, they have been deeply engaged in exploration, and their thoughts and considerations are reflected by the vivid projects in this book. For me, the book is like the author, and I feel very close to them. I wish them to continue on the road of practical exploration and academic research, constantly enrich and perfect the concept of nature in mind, and think seriously and do diligently for constructing the modern theory of Chinese garden.

Xiu Long
The Chairman of Architectural Society of China
The Former Chairman of China Construction Technology Consulting Co.,Ltd.

序三

心想、心向、心象自然
——感悟杂谈

环艺院成立于 2003 年，其创始人李存东、史丽秀都是曾经和我合作过的建筑师和室内设计师。当年他们选择转轨，投身于景观设计方向是下了很大决心的。我们之前也沟通过很多次，设想过行业的竞争和市场的认知度，也担心团队的组建、人才的引入等等。但我总体上是积极支持的，因为我们都一致认定我们的生态环境保护和社会对美好生活环境的迫切需求一定是越来越重要，急需要一大批有理想、有新思维的设计团队。今天看来，他们这个方向选对了，转向也很及时，团队不断壮大，作品频频推出，今年又更名为景观生态院，旗帜更加鲜明！

我本人的许多项目都与他们合作。换句话说，他们是我的设计团队的一部分，是景观方面的主要依托力量，从北外、北川、鄂尔多斯到济南、海口、成都，这几年更是反过来把我拉进了他们的主场，如南宁园博会、北京世园会，在他们的景观规划中做风景中的建筑。设计风景中的建筑一定要改变心态，风景是主体，建筑要融入其中，尊重环境、敬畏自然是基本的态度，而建筑本身营造开放的空间，减少能源的消耗，引导健康的行为，采用因地制宜的技术亦是十分重要的策略。在我心里，在风景中的建筑一定是匍匐于大地、沐浴着阳光、舒展着身躯、迎向着清风，换句时髦话说，应该是充满了正能量又接地气的样子。

记得存东在和我研究南宁园博会的展览时，多次提到“心象自然”这个命题。

当时，我觉得很好，但初时听成了“心想自然”，做景观设计当然要心里想着自然！存东加重了语气又说了一遍，我又理解成了“心向自然”，也没错啊，当下城里人向往自然已成大势，每逢假日，高速路上挤满下乡进山的车就是最好的证明，设计肯定要顺势而为！存东又赶忙摇摇头，用手指在桌上笔画道是“心象自然”！一下让我恍然明了，这个词用得好！因为无论是“想”还是“向”，这一静一动的心理状态都还是浅层次的，而“象”是心中自由的一种图景。“心象自然”就是在心中的一幅山水画。可以设想设计师通过观察自然、遍览风景，一定会在头脑里存入大量的影像资料，而这些资料又与其自身的诸多主观内因和客观外因叠合映射，逐渐在心中显现出一幅属于自己的内心风景。每每设计伊始，这内在的心象与外在的自然相遇，便迸发出设计的灵感。在自然环境中注入心灵的意象，完成一次主客观之间的交互，营造出一片人工干预过的、为人服务的“半自然环境”，这可能就是景观设计的本质。如此一来，设计是否得体、能否与真的自然风景协调，就要看设计师的“心象”够不够自然了。存东用这个词做主题，我想一方面是阐述他和他的团队的理念，另一方面也是设定了一个目标、一个标准，让大家明确方向，自觉自律，时常问问自己，心中的那幅山水画画得怎么样了？……

崔愷
中国工程院院士
中国建筑设计研究院总建筑师

Foreword Three

Think of in Mind, Yearn for in Mind, Inner Nature Imagery
— inspiration and miscellaneous talk

Environmental art institute was founded in 2003, the institute's founders, Li Cundong and Shi Lixiu, are architects and interior designers I once worked with. At that time, they chose to switch to the direction of landscape design is a great determination. We had communicated with each other many times before. We had imagined the competition in the industry and the recognition of the market. We were also worried about the formation of the team and the introduction of talents. But I was generally positive and supportive, because we all agreed that our ecological environment protection and society's urgent need for a better living environment must be more and more important, and a large number of design teams with ideals and new thinking were urgently needed. Today, it seems that they chose the right direction, the switch was also very timely, and the team continues to grow, the works are frequently launched. It is also renamed as the landscape ecological institute this year, and the flag is more distinct!

I have worked with them on many of my own projects. In other words, they were a part of my design team, were the main strength of the landscape aspect, from Beiwai, Beichuan, Erdos to Jinan, Haikou, Chengdu, and even pulled me in turn into their home field in recent years, such as Nanning Garden Expo, Beijing International Horticultural Exhibition, doing the architecture in the scenery in their landscape planning. Designing architecture in the scenery must require a change of mindset, that landscape is the main body, architecture should be integrated into it. Respect for the environment and revere for nature are basic attitudes. The architecture itself, that creates open spaces, reduce energy consumption, guide healthy behavior, adopt technology tailored to local conditions, is also very important strategy. In my heart, the architecture in the scenery must be creeping on the earth, bathing in the sun, stretching the body, facing the wind, in other fashionable words, it should be full of positive energy and earthy.

I still remember that Cundong mentioned the proposition of inner nature imagery for many times when we were studying the exhibition of Nanning Garden Expo. At that time, I thought it was very good, but at the beginning it sounded like "think of nature in mind".It is absolutely for landscape designers to think of nature in mind. Cundong stressed the words and said again, and I understood it as "yearn for nature in mind". It is also right, the current people who live in the city yearning for nature has become a trend. Every holiday, the highway crowded the cars to the countryside and into the mountains is the best proof, the design must take advantage of the trend! Cundong quickly shook his head again, with his finger gesticulating on the table "inner nature imagery"! He lets me suddenly be enlightened, that's a good word! Because no matter "think of" or "yearn for", the state of this quiet and moving heart is still shallow, and "inner nature imagery" is a picture of the freedom of the heart. "Inner nature imagery" is a mountains-and-waters painting in the heart. It can be assumed that the designer will store a large amount of image data in his mind by observing the nature and browsing the landscape, and these data will be superimposed and mapped with many subjective internal and objective external factors, gradually showing a picture of his own inner landscape in his heart. At the beginning of each design, the inner mind meets the outer nature and bursts out the inspiration of the design. Inject mental intention in the natural environment, complete an interaction between the subjective and the objective, create a manually intervened "semi-natural environment" to serve the people, which may be the essence of landscape design. In this way, whether the design is decent or not, whether the design can coordinate with the real natural scenery or not, it depends on whether the designer's "inner imagery" is enough natural or not. Cundong uses this word as the theme. I think on the one hand, it is to set forth the concept of him and his team, and on the other hand, it is to set a goal and a standard, so that let everyone get clear direction and be consciously self-discipline, often ask yourself, how is the landscape painting in the heart going?

Cui Kai
Academician of the Chinese Academy of Engineering
Chief Architect of China Architecture Design & Research Group

序四

有感“心象自然”

庚子年 9 月初，在内蒙古工业大学的一个学术活动中，学会存东秘书长希望我为新书《心象自然》写个序，并将书稿发来。经披阅通读，感觉此书出版不仅具有很高的专业价值和学术意义，而且在我国风景园林的工程实践方面具有较好的引领作用。

一是看到书中所选项目范围广、跨度大，完成度很高。既有小尺度的景观和建筑，又有大尺度的景区和公园，既有一般性的项目，也有众所周知的诸如国家体育场（鸟巢）、布达拉宫、园博会、世园会等重点项目，体现出一个大设计院的设计状态和设计水准。得知李存东、史丽秀二人是从 2002 年开始创办中国建筑设计研究院景观所，从无到有，建章立制，设计研究，历经十八载，积累如此多的好作品，着实不容易，非刻苦耐劳不能有此成绩。

二是这本书不但是作品的集结，更是将作者所思所悟融于创作过程中，尤其心象自然理念的提出，提升了本书的价值内涵。一般来说，谈及对景观设计的理解，大多会从功能性、生态性、文化性等影响要素层面去分析。而本书另辟蹊径，从设计师创作主体角度进行剖析，探讨如何将“眼中之竹”转化为“心中之竹”再到“手中之竹”的过程，具有方法论层面的意义。“象为自然，与心相应”，很好地道出了景观设计的创作本质。将项目按照心象、自然以及心象自然加以归类整理，从二元对立到相互统一，使我们既了解创作过程中的心路历程，也对景观设计理论研究有了新的启发。“心象自然”既是“道法自然”哲学思想具体而微的景观呈现，也是“虽有人作，宛自天开”中国传统造园思想的现代诠释，既试图拓宽创作主体的心象空间，又避免过度“以人为本”而导致的自然缺失，使主观与客观、人与自然在景观设计中更加趋于平衡，随项目的差异而表现出很好的适应性。

三是此书出版，还具有专业教育层面的意义。从专业划分看，此书突破了景观与建筑、规划的边界，所含内容看似以景观为主，实则涉及规划、建筑，甚至桥梁水系等市政领域以及雕塑标识等环境艺术领域，很多大尺度的项目已经用城市设计的方法统筹协调各要素的关系。这种广泛性和包容性是应该鼓励的，在信息化和技术进步日新月异的当下，跨领域研究和跨专业协同是很好的方法，"一专多能"已经成为趋势，"即时学习"已经成为非常重要的专业能力。因此，《心象自然》这本书不只对设计师有借鉴意义，对在读的专业学生也会有很好的启发。

期待存东和他的团队保持初心，探索理论，持续实践，在将来取得更加辉煌的业绩和成就。

是为序。

王建国

中国工程院院士

东南大学教授

Foreword Four

Reflections on Reading "Inner Nature Imagery"

Early September of the year 2020, in an academic activity of Inner Mongolia University of Technology, Secretary-General Cundong of the society hoped that I would write a preface for the new book, Inner Nature Imagery, and send the manuscript. After reading through the book, I feel that this book's publication has high professional value and academic significance and has a right leading role in the engineering practice of landscape architecture in China.

First is to see that the book's selected projects have a wide range, a large span, and a high degree of completion. There are not only small-scale landscapes and architecture but also large-scale scenic spots and parks. There are general projects notable vital projects, such as the Beijing National Stadium (Bird's Nest), Potala Palace, China Internation Garden Flower Expo, International Horticulture Exposition, reflecting the design status and level of a large design institute. It is learned that Li Cundong and Shi Lixiu founded the Landscape Architecture Institute of China Academy of Architectural in 2002. After 18 years of research, they have accumulated many good works. It is not easy for them to have such achievements without hard work.

The second is that this book is a collection of works, combining the author's thoughts and insights in the creative process. In particular, the idea of Nature in Mind is proposed, which enhances this book's value. When discussing Landscape design, most of them will analyze it from functional, ecological, cultural, and other influencing factors. However, this book takes a different approach, analyzing from the perspective of the designer's creation subject, and discussing how to transform the "bamboo in the eye" into the "bamboo in the heart" and then to the "bamboo in the hand" process, which has methodological significance. "The image is the nature, and the heart is in touch," which well expresses the creative essence of Landscape design. The projects are categorized according to mental imagery, nature, and nature in mind, from duality to mutual unity, so that we understand the mental process in the creative process and have new inspiration for the theoretical research of landscape design. "Inner nature imagery" is not only a concrete and microscopic presentation of the philosophical thinking of "Dao follow the nature," but it is also a modern interpretation of the traditional Chinese gardening thought. "Being artificial, landscape is comparable to a natural wonder." It attempts to broaden the mental image space of the subject of creation and avoids excessive "people-oriented" The resulting lack of nature has made the subjective and objective, human and nature more balanced in the landscape design, showing great adaptability with the difference of the project.

The third is that the publication of this book also has the significance of professional education. From the perspective of professional division, this book breaks through the boundaries of landscape, architecture, and planning. This kind of universality and inclusiveness should be encouraged. In the current era of rapid informatization and technological advancement, cross-field research and cross-professional collaboration are useful methods. "One specialization, multiple abilities" has become a trend, and "Instant learning" has become an essential professional ability. Therefore, the book Inner Nature Imagery is useful for designers and inspiring students who are studying.

I hope that Cundong and his team will maintain their original aspirations, explore theories, continue to practice, and achieve more brilliant performances and achievements in the future.

It's for the preface

Wang Jianguo
Academician of the Chinese Academy of Engineering
Professor of Southeast University

前言

作品后面的设计观

自然景观广漠神奇，鬼斧神工，常常令我们叹为观止。人类的设计也会在自然中绽放异彩，不时闪耀智慧的光芒。

在众多设计门类中，园林景观设计是独特的一枝，它发生在人类聚居的土地上，有人，才有园林景观设计。跟其他设计不同的是，只有人还不行，园林景观设计离不开自然。它是人造自然，始于自然，归于自然，甚至所用的造景材料都是自然的，花、草、树、水、石、土……
原本，人是自然的一部分，景观设计应该尊重自然、顺应自然，追求天人合一、人与天调。

然而，大量人工环境中片面追求“以人为本”，只考虑人，不考虑自然的设计比比皆是。我们有必要回归本源，再次思考人和自然的关系。
其实，人和自然，是园林景观设计永恒的主题，园林景观设计的核心，在于平衡人和自然的关系。园林景观设计师是所有建成景观的始作俑者，我们必须对主观与客观、人与自然的关系有所回应。

于是，经过多年的实践和思考，我们提出了“心象自然”的设计观。
心象：内心意象的主观呈现。
自然：自然而然的客观存在。

心象自然，
是一种观念：设计要平衡人和自然的关系，满足二者的需求。
是一种态度：人是自然的一部分，要以人为本，更要以自然为本。
是一种方法：将自然内化于心，再由心外化于景观。
是一种追求：抒发内心对自然的感受，象为自然，与心相应。

“心象自然”就是要平衡人和自然的关系，不是只考虑人类自身的需求。我们既要延续自然界的生态准则，又要保持人类文明持续健康发展。我们追求的“心象自然”是一种境界，用自然反映人性的真实，用内心体悟自然的美好。我们表达的，是与自然相通的内心世界。

李存东

Preface

The Design Concept behind Projects

The natural landscape is so vast and impressed that could easily take our breath away, but human designs can also shine wisely in nature.

In numerous design categories, landscape design is a unique one. It happens on the land where human live. Human is the start point of landscape design; however, unlike other designs, it cannot be built apart from nature. Although, landscape design is creating artificial nature, it starts from nature and ends to nature. Therefore, this design should present full respect to nature, adapt nature, and even pursuit of the human-nature harmony. Nevertheless, a mass of artificial designs blindly seeks after "humanist", and against the law of nature. We, as designers, must return to the origin, and reconsider the relationship between human and nature.

In fact, human and nature are landscape designs' eternal theme and design core which lead to the method balancing these two ends of the scale. Landscape designers are the initiator of all built landscape projects, and we must respond to the both subjective and objective human-nature relationship.

Therefore, after years of practice and thinking, we provide a design concept of "Inner Nature Imagery".
Inner Expression: Subjective expression of inner image.
Nature: Natural objective existence.

Inner Nature Imagery,
is a concept: Design is a tool to balance the relationship between human and nature while satisfying requirements from both of them.
is an attitude: Human is a part of the nature. It is important to meet the needs of human, but following the Nature-Base principle is the original intention.
Is a method: Internalized nature into heart, and externalize it into the landscape design.
Is a pursuit: Expressed the inner feelings of nature, and echo to the heart.

"Inner Nature Imagery" concept is not only to serve human, but to balance the weight of human and nature. We hope to continue the ecological principle, and as meanwhile, keep human civilization developing healthily. We pursuit "Inner Nature Imagery" as a realm to reflect the truth of humanity and realize the beauty of nature. In general, the main point that we are representing is the inner world that connecting nature.

Li Cundong

目录

Contents

心象篇

Inner Imagery

心象

无论景观作品最后呈现出什么样的形态，背后都是通过园林景观设计师的设计来作为实施依据的。影响设计的因素有很多，最重要的还是设计师内心意象的主观呈现。

设计师的内心意象有时会强于自然环境对他的影响，尤其在以人工环境为主的城市空间中。这时候设计师对大自然的呼应少一些，而内心中或理性或感性的因素慢慢地外化出来。

理性的外化是基于对人的需求的理解和回应，在城市客厅、交通枢纽、旧城改造等场景中，我们尽可能还原出人们活动其中的画面感，用分析问题、解决问题的理性思维去主导设计。

感性的外化会更复杂、更不好把握，它往往随情境而变化。在一些功能要求不是太严格的场地中，我们的心性会更自由、更放松，更容易感性地创作。作为受过专业训练的设计师，我们愿意尝试对艺术的追求、对文化的探索。我们愿意对形体进行研究，对水、石、小品等进行塑造，让景观呈现出的形态有趣一些，设计感更强一些。我们会在一些民族地区，把当地的民族文化带到设计中，体现差异化和地域认同。我们也会把主流文化的传统语素加以提炼，寻求古韵新作，探索传统与现世的共生。我们对城市更新中场地的记忆加以保留，尽可能体现历史文脉的痕迹。我们也尝试让中国园林走出国门，把园林文化传播出去，使中国优秀的传统文化得以宏扬……

Inner Imagery

Landscape designer is the main decisive factor of how the project will present, and in these tons of elements that influencing design quality, the subjective presentation of designers' inner intention is the most crucial one.

Sometimes, designer's inner intention is stronger than the affection from nature, especially in the urban space where dominated by artificial environment. At this moment, designers should react less to the nature in order to gently express their inner emotional or rational feelings. The externalization of rational feelings is based on the understanding and response to human needs. In urban living rooms, transportation hubs, old city reconstruction, and other scenes, we try to restore the scenario with human activities, and then apply rational thinking to analyze while solving problems to lead the design. The emotion externalization is more complex, mysterious, and inconstant in different situations. Human feels more comfortable to create in the place that require less functions, and as trained designers, we would like to explore more on art and culture. We are willing to study water, stone, and outdoor furniture's form shaping and present their amusing aspect with stronger sense of design. In some ethnic areas, we integrate local culture into the design to represent the differentiation and their specific regional identity. It is also our job to refine the traditional elements of the dominant culture. In order to finding new ways to express tradition, we hope to explore the symbiosis between tradition and the present. We maintain and protect the site memory in the developed city to reflect historical traces, and we also try hard to send Chinese landscape to the world stage, and take this opportunity to promote Chinese culture.

第一章

理性的外化

Chapter One

Externalization of Reason

一　城市客厅
City Parlor

国家体育场鸟巢
National Stadium, the Bird's Nest

赛前赛后

2008年奥运会在中国人心中的分量是举足轻重的，作为国家主体育场的鸟巢，不只在奥运期间牵动了亿万人的心，在赛后运营的十多年时间里，也一直像是北京的城市客厅，每天都会接待大量的参观者，还会举办各种类型的活动。作为活动的必经之地，鸟巢外围14.75公顷的景观区域显得尤为重要，这个场地必须既能适应平时游客的观赏游憩，又能在大型活动时保证十几万人短时间的集散和安检。也就是说，在这样一个城市级的大客厅里，我们的设计既要保持轻松浪漫的格调，又要更加理性地满足各类人群的使用需求。

树根与树

体育场的景观设计开始于2004年，虽然时过境迁，但当你今天来到鸟巢，穿过周边的绿荫和草地，沿着由石英岩铺砌的不规则网状道路缓缓向上，欣赏景致的同时逐渐接近鸟巢，你仍然能够感受到当初的设计意图，景观的秩序延续了体育场的结构肌理并与之合二为一，如同树根与树的关系。当你接近鸟巢回望周边，不经意间会感受到场地的抬高，也能更好地体会到奥林匹克中心区整体的氛围。

Before and After the Game

The 2008 Olympic Games are of great importance to the Chinese people. As the main national stadium, the Bird's Nest not only touched the hearts of millions of people during the Olympic Games, but also served as the city parlor of Beijing during the ten years after the games, receiving a large number of visitors every day and holding various kinds of activities. As the only way which must be passed, the landscape area of 14.75 hectares outside the Bird's Nest is particularly important. This site not only must be suitable for tourists' appreciation and recreation at ordinary times, but also ensure that more than 100,000 people can gather and disperse in a short time and go through security checks during large-scale activities. That is to say, in such a city-level parlor, our design not only should maintain a relaxed and romantic style, but also more rational to meet the use needs of all kinds of people.

Roots and Trees

Landscape design for the stadium began from 2004. Although times flies, when coming to the Bird's Nest today, through the surrounding shade and grassplot, you slowly go up to the irregular network roads paved with quartzite. You can still feel the original design intention, while enjoying the view and approaching the nest: the order of the landscape continues the structure and texture of the stadium and unit as one with it, like the roots of trees. As you approach the Bird's Nest and look back at the surrounding area, you can't help but feel the elevation of the field and can also better appreciate the overall atmosphere of the Olympic central area.

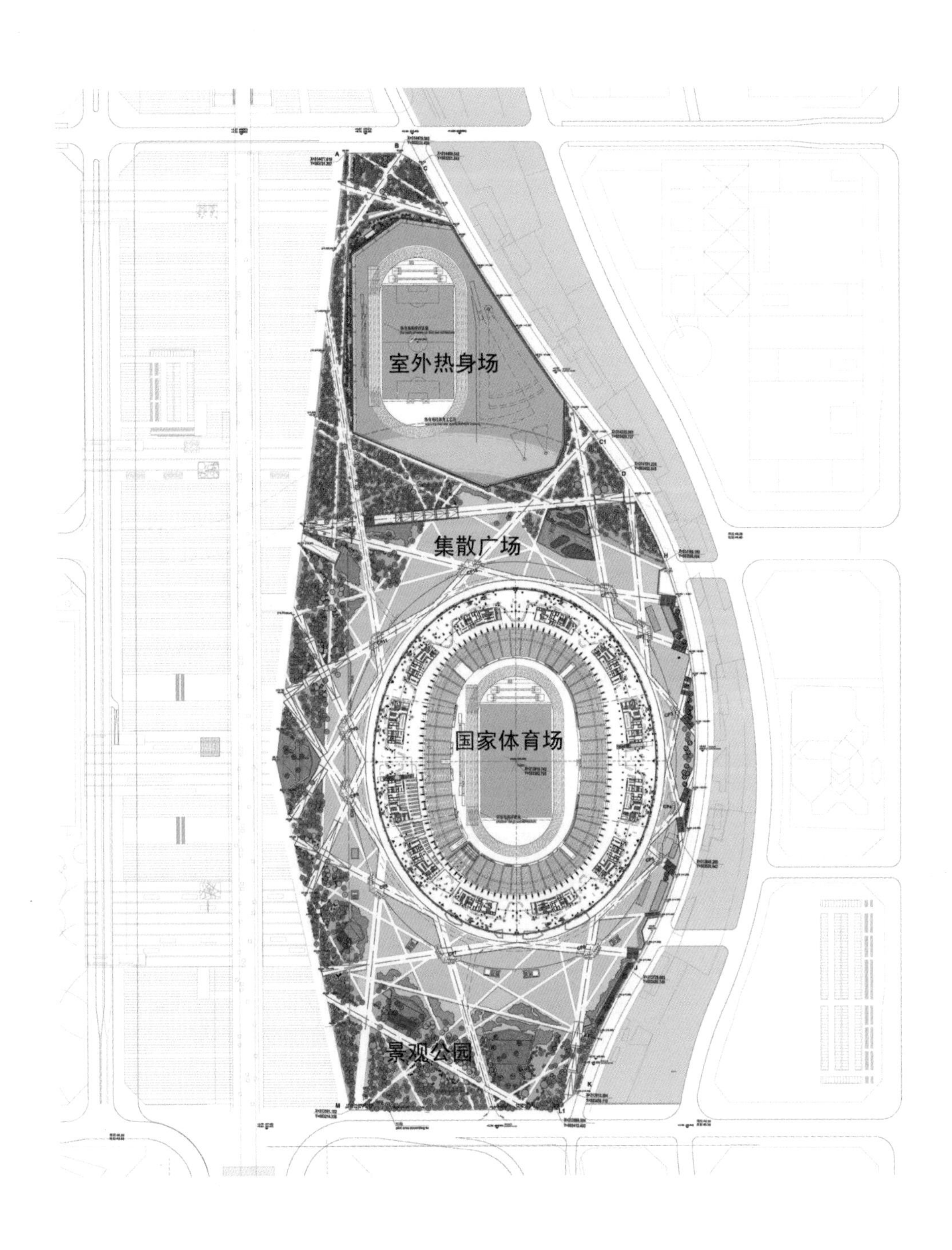
室外热身场
集散广场
国家体育场
景观公园

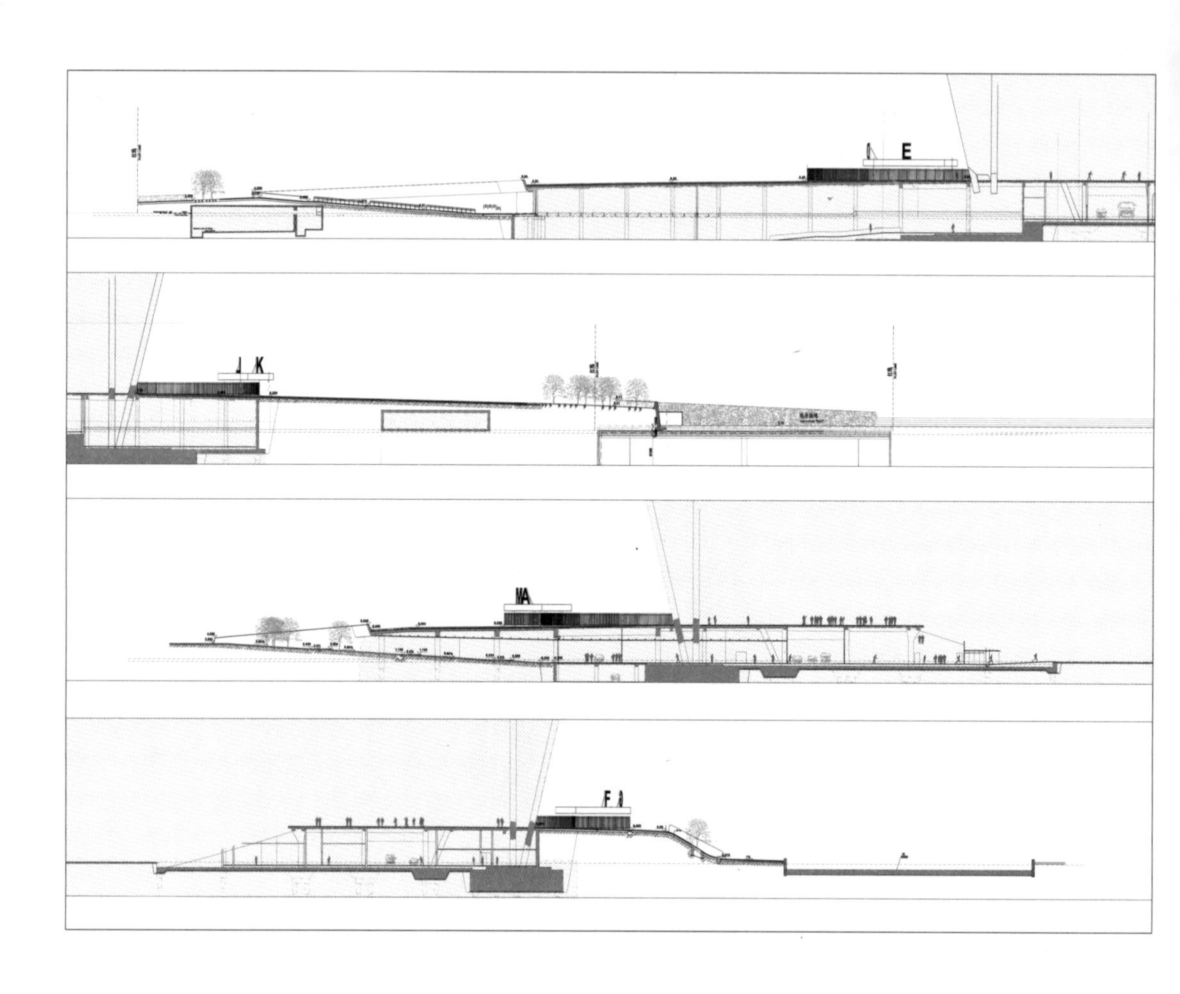

功能优先

体育场周边的功能有它的特殊性，满足集散是首要的，越接近体育场空间越开阔，12个主要出入口与集散广场和9米宽的主路相连，绿地接近出入口的地方也作为应急疏散绿地，使得场地的景观肌理更加丰富。安检是体育场的又一个重要功能，安检围栏设置在离体育场约20米的位置，从而保证进入安检后的开阔空间。无障碍设计是考虑的重点，9米宽的主要道路能够满足轮椅和视力残疾者的使用，3米宽的次要道路满足视力残疾者使用，有高差变化处和不适合轮椅通行的道路两端均布置了提示标志，在9处安检口设置了18个无障碍检票通道，在南北两处室外下沉售票点设计了无障碍坡道和残疾人售票窗口，场地内饮水台和公用电话也都能适应残疾人的使用。

Function Priority

The function around the stadium has its particularity, and satisfying the gather and dispersing function is paramount. The closer getting to the stadium, the widener space is. The 12 main entrances and exits are connected to the gather and dispersing square and the 9-meter-wide main road. The green space near the entrances and exits is also used as emergency evacuation green space, which makes the landscape texture of the site richer. Security check is another important function of the stadium. The security fence is located about 20 meters from the stadium to ensure the open space after security checks. Barrier free design is a key consideration. The 9-meter-wide main road is suitable for people with wheelchair and visual disabilities, and the 3-meter-wide secondary road is suitable for people with visual disabilities. There are warning signs at both ends of the road where there are elevation variations or the road is not suitable for wheelchairs. 18 barrier free check-in channels are set up at 9 security checkpoints and barrier free ramps and ticket windows for the disabled are designed near the outdoor sunken ticket offices in the north and south directions. The drinking water station and public telephone are also suitable for the disabled.

K
K
L

做深做细

在这样一个重要的场所，景观设计需要考虑的因素远远超出了一般设计。坡形场地上有多个通向地下的人行和车行出入口、疏散楼梯开口、通风口以及复杂的台阶、坡道、挡墙、排水沟等需要处理；体育赛事需要的售票处、公共卫生间、安检围栏和配有太阳能装置的安检棚等需要特殊设计；体现中国特色和地域特点的种植设计以及满足奥运期间开花的特色植物品种需要认真研究；特色的碎拼铺装与建筑的交接关系需要不断编织；各种功能设施，包括室外座椅、垃圾箱、背景音乐、应急广播、安保监控、照明系统、喷灌系统、标识系统、直饮水台、ATM机、公共电话、旗杆等需要不断配置……太多太多的工作需要研究、交流、汇报、修改……细细回味，就是这些一个个的细节，填充了我们4年多辛勤设计的日日夜夜……

Do Deep and Fine

In such an important site, landscape design needs to consider much more factors than the general design. There are many entrances and exits for pedestrian and vehicle leading to the underground, openings of evacuation stairs, vents, as well as complicated steps, ramps, retaining walls, and drainage ditches on the sloping fields that need to be treated; ticket offices, public washrooms, security barriers and security sheds with solar panels for sporting events need to be specially designed; planting design that reflects Chinese characteristics and regional characteristics and the special plant varieties that can bloom during the Olympic Games are necessary to study carefully; the characteristic patchwork pavement and the handover relationship of architecture need to be woven constantly; various functional facilities, including outdoor seats, garbage cans, background music, emergency broadcast, security monitoring, lighting system, sprinkling irrigation system, sign system, direct drinking water station, ATM, public telephone, flagpole, etc. need constant configuration. Too much work needs to be researched, communicated, reported, and revised. Enjoying them carefully, it is these details that fill our more than four years of hard design day and night.

入口
Entrance

拉萨布达拉宫周边及宗角禄康公园

Surrounding of the Potala Palace in Lhasa and the Zongjiao Lukang Park

神圣的世界文化遗产

布达拉宫被称为“世界屋脊明珠”，是松赞干布为迎娶尼泊尔尺尊公主和我国唐朝文成公主而兴建的大型宫殿，始建于公元7世纪，至今已有1300多年的历史。17世纪重建后，它成为历代达赖喇嘛的冬宫居所，同时是西藏政教合一的统治中心。“筑一城以夸后世”，布达拉宫依其磅礴的气势和独特的建筑风格，在中国乃至世界古代建筑史上都占有极其重要的地位，是藏传佛教的圣地，每年至此的朝圣者及游客不计其数，1994年12月布达拉宫入选世界文化遗产名录。

不相称的遗产环境

如此神圣的文化遗产却有着不相称的遗产环境。布达拉宫周边挤满了体量过大、布局呆板的现代建筑，其东、北、西三侧道路边布满了“一层皮”式的 3 层沿街商业建筑，这些建筑将红山及布达拉宫包围得密不透风，市民无法看到，环境问题异常突出。

在布达拉宫北侧山脚下，当年达赖五世取土建宫时留下龙王潭，后围绕它形成宗角禄康公园。因被周边商店所围，缺少维护而逐渐荒废，少有人来。

Sacred World Cultural Heritage

The Potala Palace, known as the "pearl on the roof of the world", is a large palace built by Songtsen Gampo, for his marriage to Nepal's Princess Bhrikuti and China's Tang Dynasty Princess Wencheng. It was built in the 7th century and has a history of more than 1300 years so far. Rebuilt in the 17th century, it became the winter palace residence of successive dynasties of Dalai Lamas and also the ruling center of unification of the state and the church in Tibet. "Build a city for posterity to brag about." The Potala Palace, depending on its magnificent momentum and unique architectural style, occupies an extremely important position in the history of ancient architecture in China and even the world. It is the holy land of Tibetan Buddhism, attracting countless pilgrims and tourists every year. In December 1994, the Potala Palace was included in the world cultural heritage list.

Unfitting Heritage Environment

Such a sacred cultural heritage has an unfitting heritage environment. The Potala Palace is surrounded tightly with oversized, rigid modern buildings, and its east, north, and west side of the road is covered with three-story commercial buildings which have the same architecture style. These buildings surrounded tightly the Red Mountain and the Potala Palace, so that not only the public could not see the Potala Palace, but also the environmental problems are particularly prominent.

At the foot of the hill to the north of the Potala Palace, when the fifth Dalai Lama took earth to build his palace, Longwangtan was left behind, and then formed the Zongjiao Lukang Park around it. Surrounded by the shops, the Zongjiao Lukang Park was lack of maintenance and gradually abandoned, so that few people came.

市民的日常需求

转经、喝酥油茶、过林卡……是拉萨市民的日常活动，尤其是转经，拉萨转经道总有人在转，他们身上带着念珠，一边不停地摇转手中的转经筒，一边不停地摇转路边的经筒。围绕布达拉宫转经的过程中，他们要买些日用品、他们会停下来喝酥油茶、到绿地中席地而坐，过林卡……，这些活动需要有一个好的环境去支撑。布达拉宫周边的转经道需要加以整治，宗角禄康公园的功能需要梳理、环境需要提升。

还给市民的城市客厅

还给世界文化遗产一个适宜的背景环境、还给市民一个舒适的室外空间，这是布达拉宫周边及宗角禄康公园最直接的需求。于是“拆”便成为首要策略，拆掉周边不适宜的商业建筑，打开布达拉宫周边的视线；拆除围墙和摊位，使布达拉宫周边成为一个开放的城市空间。有了“拆”自然就要“保”，要保护好现状的文物建筑、成片的树林、已有的水面以及拉萨市民的生活习惯；在此基础上再采用“梳”的策略，就是要梳理“拆”与“保”之后的整体环境，梳理场地的特征并植入相应的功能，包括格桑花广场、甜茶馆、游船码头、健身器械等，满足市民的活动需求，还给市民一个自然轻松的城市客厅。

Daily Needs of the Public

Turn scripture, drink buttered tea, celebrate the Linka Festival…they are the daily activities of Lhasa residents, especially turning scripture. There are always some people turning scripture in Lhasa. They wear rosary beads, and keep shaking the prayer wheel in their hands, while constantly shaking the prayer wheel on the side of the road. When turning scripture around the Potala Palace, they would buy some daily necessities, stop to drink butter tea, sit on the ground in the green space, and celebrate the Linka Festival. These activities need to be supported by a good environment. The turning scripture roads around the Potala Palace need to be regulated, the functions of Zongjiao Lukang Park need to be sorted out, and the environment needs to be improved.

City Parlor Returned to the Citizens

Returning a suitable background to the world cultural heritage and a comfortable outdoor space to the public is the most direct demand of the Potala Palace and Zongjiao Lukang Park. Thus, "dismantling" becomes the primary strategy. Dismantle the inappropriate commercial buildings around the Potala Palace and open the view around the Potala Palace; dismantle the walls and stalls, to make the Potala Palace an open urban space. Preserving is another necessary aspect except dismantling. Preserve the existing cultural relics, the forests, the existing water, and the living habits of Lhasa residents; on this basis, adopt the strategy of "combing", to comb out the overall environment after "dismantling" and "preserving", to comb out the features of the site and implant corresponding functions, including Gesanghua Square, sweet teahouse, boat wharf, fitness equipment, etc., to meet the needs of the public activities and give the public a natural and relaxed city parlor.

二 城市枢纽
City Hub

苏州火车站
Suzhou Railway Station

隐于城市的尺度

苏州，一座举世闻名的“天堂”城市，在这里“水城人家尽枕河，鱼米之乡遍书声”，无论纵横交错的小桥水网、清新恬淡的粉墙黛瓦，还是缥缈朦胧的吴门烟水、暗香浮动的花光月影，无处不在地展示着这片富有才华的江南锦绣之地的品位和气韵。一直以来，“小桥—流水—人家”的宜人尺度成就了苏州特有的城市气质和文化底蕴。

在这样气质的城市中，而且是在苏州古城护城河西北岸边构建超尺度的火车站建筑，无疑会对老城乃至整个城市的尺度产生强烈的冲击，建筑设计用重复的菱形体空间分解了庞大的建筑屋顶，用大开口景墙分解建筑横纵尺度，使体量庞大的建筑轻轻地隐于城市的尺度之中。

大尺度的建筑必然有相适应尺度的站前广场，为了能将广场融入建筑、融入城市，需要用更加细腻、更加适宜的手法，让广场空间转化成建筑、城市与人三者尺度过渡的粘合剂，这必然要对广场空间进行理性的分解，使广场既要与建筑尺度协调，又要隐于城市的尺度之中。

Hidden in the Scale of the City

Suzhou, is a world-famous "paradise" city, "a place where the houses there are built on the river, and an abundant place full of the sound of books". No matter crisscrossing small bridge and water network, fresh and light white wall and black tiles, or ethereal and hazy environment of Wumen, flower and moonlight shadow floating with fragrance, everywhere is showing the taste and charm of this piece of gifted place, south of the Yangtze River. All this time, the pleasant scale of "small bridge, flowing water and human houses" has contributed to the unique urban temperament and cultural deposits of Suzhou.

In a city with such a temperament and on the northwest bank of Suzhou's ancient moat, the super-scale railway station construction will undoubtedly have a strong impact on the scale of the old city and even the whole city. The architectural design decomposed the huge building roof with repeated rhomboid space, and decomposed the horizontal and vertical scale of the building with large opening landscape walls, so that the massive building is gently hidden in the scale of the city.

Large scale buildings must have appropriate scale of station square. In order to integrate the square into the architecture and the city, it needs to use a more delicate and appropriate method to transform the square space into a binder for the transition between architecture, city and people. It is necessary to rationally decompose the square space, so that the square should not only coordinate with the architectural scale, but also hide in the scale of the city.

归于广场的功能

广场空间的分解要源于广场的功能属性和主要人流方向的理性分析。广场要重点解决整个城市人流进站、出站、步行穿越、地上地下交叉、水陆交叉等各类人流以及自行车、出租车、长途汽车、公交车等各类车流的功能要求。通过分析，南北广场的主要交通流线分别呈现“一”字形和“十”字形趋势，这个趋势将统领广场的功能分区和空间划分。

化于院落的形式

明确了功能，形式便成了解决空间尺度的主要途径。通过分析后发现，广场中主要人流以外的空间是解决广场尺度问题的重要部位，于是借鉴苏州园林的院落空间处理方式，在南北广场分别打造了“弓”字形和“U”字形院落，形成“院落广场”的空间特质，将大尺度的广场院落化处理，突出大小院落空间与广场之间的转换串联，实现建筑、广场、城市与人之间尺度的有机协调，从而形成富有个性的新院落形式。

Function Attributed to the Square

The decomposition of the square space comes from the rational analysis of the functional property of the square and the main flow of people direction. The square should focus on solving the functional requirements, such as people in the whole city entering and exiting the stations, walking and passing through, crossing ground and underground, crossing water and land, as well as various kinds of people flow and traffic flow such as bicycle, taxi, coach and bus. Through the analysis, the main traffic flow lines of the north and the south square respectively show the trend of "一" and "十", which will dominate the functional division and space division of the square.

Form Embodied in the Courtyard

Once the function is defined, the form becomes the main way to solve the spatial scale.Through analysis, it is found that the space outside the main flow of people in the square is an important part to solve the problem of square scale. Therefore, the treatment of courtyard space in Suzhou gardens is learned from. The courtyards in "弓" type and "U" type are respectively built in the north and the south square, forming the space characteristic of "courtyard square". The large-scale square is divided into fractal courtyard, highlighting the conversion and series connection between the courtyard space and the square. Achieve the organic coordination of the scale between the building, square, city and people, so as to form a new courtyard forms with rich individual character.

成于苏式的风格

有了形式，院落风格的确定就至关重要了。整体院落从苏州古典园林中汲取营养，并用与建筑风格协调的材料和颜色，借鉴苏州当地能工巧匠的工艺，与建筑一起共同体现一种“生于斯，长于斯”的苏式新风格。建成后的效果是成功的，既有效地解决了火车站广场的所有功能，所形成的院落气质又隐含地体现了苏州园林的特征。

Style Succeed in the Su Style

When the form is confirmed, the determination of the courtyard style is crucial. The whole courtyard draws nourishment from the classical gardens in Suzhou, with materials and colors that harmonize with the architectural style, drawing lessons from the craftsmanship of local skilled craftsmen in Suzhou, which together with the architecture embodies a new style of Suzhou that is "born in here, grows in here". The effect after construction is successful, which not only solves all functions of the railway station square effectively, but also reflects implicitly the characteristics of Suzhou gardens by the temperament of the courtyard.

拉萨火车站

Lhasa Railway Station

天路明珠

顺着潺潺流动的拉萨河远远望去，在那山水之间闪烁着一颗红白相间的晶体，这就是拉萨火车站，一座被誉为“天路”的青藏铁路线上的标志性建筑。拉萨火车站位于拉萨市南部的柳吾新区，与市区隔河相望，站房周围是连绵的山脉，场地宽阔而平坦，是西藏面向外界的重要门户。

漏斗塑形

由于场地南高北低，周边交通复杂多样，建筑向外视线开阔，从建筑向外看，远山、中树、近场浑然一体，于是广场由室内向室外延伸成为图底的必然选择。顺着视线方向和人流方向，将建筑入口和北侧道路交叉口连接，形成一个漏斗式下沉中心广场，控制整体空间秩序，而两侧的长途汽车站、出租车站、公交车站和公共停车场利用这种图底空间高效、便捷地分散着进出站的人流。

Bright Pearl in Heaven Road

Seen from a distance along the flowing Lhasa River, a red and white crystal gleams between the mountains and rivers. This is Lhasa railway station, a landmark building on the Qinghai-Tibet railway line known as the "heaven road". The Lhasa railway station is located in the Liuwu district, in the south of Lhasa city, facing the urban area across the river. The station houses are surrounded by continuous mountains, and the site is wide and flat. Lhasa railway station is an important gateway of Tibet to the outside world.

Funnel Shape

As the site is high in the south and low in the north and the surrounding traffic is complex and diverse, the building has a wide view towards outside. Looking out of the building, the distant mountains, middle trees and near field are all in one. So the square's extension from the indoor to the outdoor became the inevitable choice of the bottom of the picture. Following the line of sight and the flow of people, the entrance of the building is connected with the road intersection on the north side, forming a funnel-shape sunken central square to control the overall spatial order. And the long-distance bus station, taxi stand, bus stop and public parking on both sides use this kind of bottom space to disperse the flow of people in and out of the station efficiently and conveniently.

点状插绿

场地交通流线较多，缺乏原生树种，大面积绿化难以实现，为了让场地充满生机，只能点状插绿。在人们匆忙的脚步下，形态各异的地形岛、造型别致的景观灯、朴实无华的树池座凳、恰到好处的指示牌、明暗交错的铺装纹理以及西侧八字形树、林荫小径等景观随时会映入眼帘，在不经意间将人们带入一个神奇的旅游世界。

Intersperse Plants in Dot Shape

There are many traffic lines in the site, and it is difficult to achieve large area greening due to lack of native tree species. In order to make the site full of vitality, the only way is to intersperse plants in dot shape. In the hurried footsteps of people, different forms of terrain island, unique shape of the landscape lights, simple tree pool sitting stool, the suitable indicators, chiaroscuro pavement texture the trees in eight-shape in the west, tree-lined trails and the other landscape will come into view at any time, inadvertently bring you into a magical world of tourism.

三 城市更新
Urban Renewal

阜成门内大街
Fuchengmen Nei Dajie

老舍笔下的最美大街

《骆驼祥子》中，祥子对虎妞说："这儿什么都有，有御河、有故宫的角楼、有景山、有北海、有白塔、有金鳌玉蝀桥、有团城、有红墙、有图书馆、有大号的石狮子，多美，多漂亮。" 地地道道的北京本土大作家老舍先生笔下描绘的这条"什么都有"的最美大街，就是位于北京市西城区的阜成门内大街。

这条形成自元大都时期，距今已有700余年历史的京城老街，如今仍然作为城市交通次干线承担着连接老城东、西城区的重要职能，熙熙攘攘、车水马龙。在以煤炭作为主要生产生活燃料的年代里，供给北京城的煤炭皆由驼队载着，从京西门头沟自阜内大街运入内城。"煤"与"梅"同音，百姓在瓮城门洞内镌刻梅花一束作为阜成门的专属LOGO。北风呼号，漫天皆白之时，烘炉四周之人皆赞"阜成梅花报暖春"。

The Most Beautiful Street in Lao She's Book

In the Camel Xiangzi, Xiangzi told Huniu that everything here is complete, including royal river, turret of the Palace Museum, landscape hill, the North Sea, the White Tower, the Jinao Yudong Bridge, round city, red wall, library, large stone lion, how beautiful here is. The most beautiful street with "everything" in the book of Mr. Laoshe, a famous local writer in Beijing, is Fuchengmen Nei Dajie in Xicheng District, Beijing.

Formed in Yuan Dynasty, the old street of Beijing has a history of more than 700 years. Today, it still plays an important role in connecting the east and west parts of the old city as an urban traffic trunk line, bustling with traffic. In the years when coal was the main fuel for production and living, the coal supplied to Beijing was carried by caravans from the Jingxi Mentougou to the inner city from the Funei Street. As "coal" has the same sound as "mei" in Chinese, people engraved a bunch of plum blossom in Wengcheng Gate as the exclusive LOGO of Fucheng Gate. When the north wind howls and the whole city is white, people sit around the stove and praise "plum in Fucheng reports a warm spring".

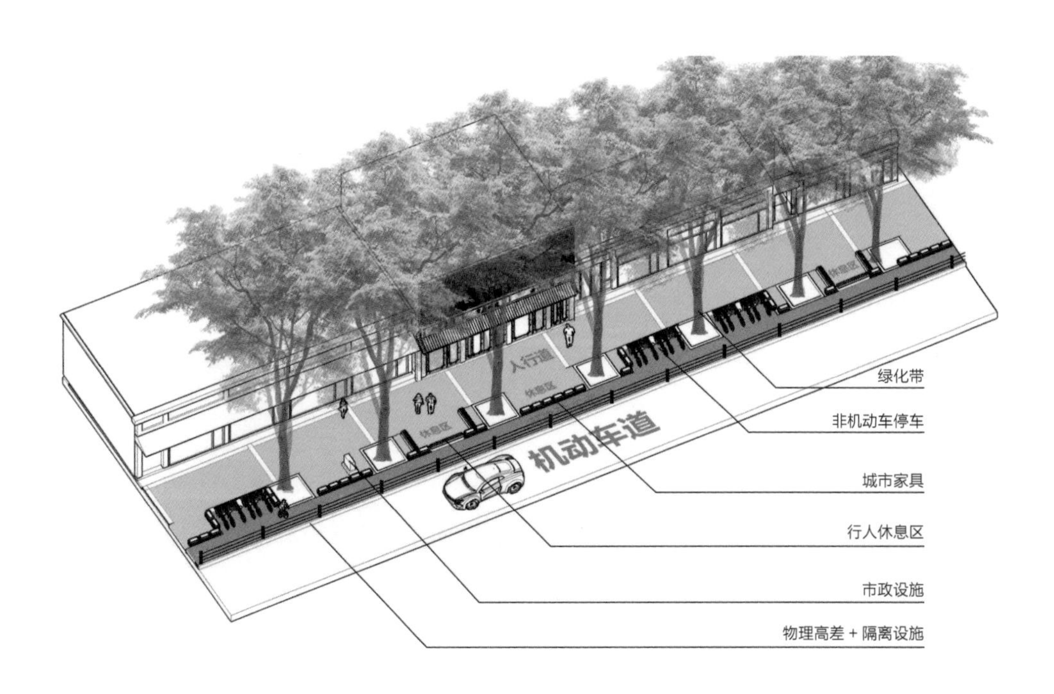

奖杯奖牌一条街

700年来，人们生活的环境发生了巨大的变化，城门拆了，煤窑关了，砂石土路变成了宽阔的柏油马路，驼队板车变成了小汽车、自行车。旧时大街两侧鳞次栉比的四合院落，一些因金融街片区的横空出世而被高楼大厦所取代，剩下的则利用临街的有利条件做起了小买卖，阜内大街也成为京城知名的“奖杯奖牌一条街”。街道空间的平衡被打破，街道的尺度越来越压抑。老街的韵味被形式各异、时髦绚丽的落地橱窗、霓虹店招牌冲淡，城市公共空间随着违法建设、机动车违规停靠现象的增多和城市设施的无序摆放被严重侵占。

对于这座古老的城市，我们过于草率地增加了太多的乱象，又过于草率地舍弃掉花费百年时间积攒的印记和底蕴。因此在阜内大街的改造过程中，我们沿着这条680米长的老街来回走了不下百次，品味着这条老街上的每一处细节，慎重地做出加法和减法。

Cup and Medal Street

Over the past 700 years, great changes have taken place in people's living environment. The gates have been dismantled, the coal kilns have been closed, the gravel and dirt roads have become wide asphalt roads, and the caravans have become cars and bicycles. In the old days, there were rows upon rows of courtyard houses on both sides of the main street. Nowadays, some of them were replaced by high-rise buildings due to the appearance of the financial street area, while others started small businesses by taking advantage of the advantages of the street. Funei Street has also become a famous "cup and medal street" in Beijing. The balance of the street space is broken, and the scale of the street is more and more depressed. The lasting appeal of the old street is diluted by the various forms of modern gorgeous french window and neon lights brand. Along with the illegal construction, the increase of illegal parking of motor vehicles and the disordered arrangement of urban facilities, the urban public space has been seriously invaded.

For this ancient city, people have excessively curtly added too much chaotic phenomena, and have excessively curtly discarded the marks and deposits that have taken a hundred years to accumulate. So in the renovation process of Funei Street, we walked back and forth along this 680-meter-long old street no less than 100 times, savoring every detail of this old street, carefully making addition and subtraction.

对城市做加减法

我们对违法建设和不合理占用公共空间的围栏进行了拆除，将较大型的变配电设备改移至室内，增加了更多的公共休憩交往空间和功能使用空间。

我们基于实地测算的交通流量数据，合并了相距不足百米的临近公交站点，减少了公交车进出站对道路交通的影响；根据街道实际使用情况，打破了原有的路板格局，通过局部段落非机动车道内绕的方式，搭配定制的机非隔离护栏，有效减少了停车占用非机动车道、人行道的现象。

我们保留了大街上所有的现状大树，并根据每棵树不同的生长情况量身定制了周边方案。另外，根据街道现状树栽植方式和种类，结合以“梅”为特色的文化内涵，新补植了各类乔木、灌木百余株，将原本灰色单调、日晒雨淋的“广场式”街道，变成了绿意盎然、树影斑驳的林荫大道。

Add and Subtract Cities

We removed fences that were illegally constructed and occupied public spaces unreasonably, moved larger transformers and distribution equipment indoors, and added more public rest and interaction spaces and functional use spaces.

Based on the traffic flow data measured in the field, we combined the adjacent bus stations less than 100 meters apart to reduce the impact of bus arrivals and departures on road traffic; according to the actual situation of the street, broke the original pattern of the road board and combine the customized non-motorized isolation guardrail with the the way of the internally installed non-motorized lane in some local sections, effectively reducing the phenomenon that parking occupies non-motorized lane and pedestrian footpath.

We have kept all the existing trees on the street, and tailored the surrounding plan according to the growth of each tree. In addition, according to planting methods and types of the current situation of the street, combined with the cultural connotation of the characteristics of "mei", more than 100 new trees and shrubs were planted. The "square" streets, which used to be dull and exposed to the sun and rain, were turned into tree-lined boulevards full of green and mottled shadows.

为了减少街道上线杆的数量，我们创新性地提出了多杆合一的方案，在2个月的时间内跑遍了城管、公安、交管、电网、公交等10余个部门和单位，征询他们的意见，得到了广泛的支持和积极的配合。最终，统合照明、电力、监控、交通指示等多项功能，将街道上原有183根各类线杆标牌减少到55根，整理优化方案按期实施落地。

在小品设施的设计中，注重传统工艺的运用、整体色调的控制和特色文化的融入。采用统一的深灰色调，让各种栏杆发挥作用的同时，消隐在老城青砖灰瓦的环境中；在行人能够近距离接触到的城市家具上，适当增加梅花纹样和浅浮雕图案，通过细节彰显景观品质。

对白塔寺这样的重点文物保护单位周边环境进行改造提升时，我们本着对文化遗产的敬畏与尊重，多次向文物主管单位和研究北京传统建筑文化及建造工艺的专家请教，在解决现存安全隐患和无障碍设施需求的前提下，融入了传统建造材料和工艺，进一步烘托和弘扬了历史文化街区的气氛。

这里的黎明热热闹闹

阜内大街是首都老城区内的“重要通道”，由于其区位的重要性和特殊性，对施工的组织、管理也提出了较高的要求。施工时段是凌晨0:00~5:00，凌晨开挖的基槽早上就要恢复，凌晨铺砌的道路早上就要投入使用，凌晨需要多少料就从料场运出多少，盈余的部分必须在黎明到来之前运回料场码放整齐。严苛又特殊的施工安排让管理方、建设方和设计方时常“深更半夜里工地相会，这里的黎明热热闹闹”。

In order to reduce the number of poles on the street, we creatively put forward a multi-pole scheme. Within two months, we went to more than 10 departments and units, including urban management, public security, traffic management, power grid and public transportation, to ask for their opinions, and got extensive support and active cooperation. Finally, it integrated lighting, electricity, monitoring, traffic instructions and other functions, reduced the original 183 signs of various kinds of wire poles on the street to 55 signs, and sorted out and optimized the plan and implemented on schedule.

In the design of landscape decoration, attention is paid to the application of traditional techniques, the control of overall tone and the integration of characteristic culture. Adopt unified dark gray tone, let all sorts of balustrades play a role at the same time, and disappear in the environment of old city's black brick and gray tile. On the city furniture that pedestrians can get close to, add plum blossom pattern and bas-relief pattern appropriately, highlight the landscape quality through the details.

When transforming and upgrading the surrounding environment of the key cultural relics protection units like the Baita Temple, we held a sense of awe and respect for cultural heritage, and consulted the unit in charge of cultural relics and experts who studied Beijing's traditional architectural culture and construction techniques for many times. On the premise of solving the existing security risks and the need for barrier-free facilities, the traditional construction materials and techniques are integrated to further enhance and promote the atmosphere of the historic and cultural district.

Dawn Here is Buzzing with Excitement

Funei Street is an important channel in the old city of the capital. Due to the importance and particularity of its location, the construction organization and management also put forward higher requirements. The construction period is from 0:00 to 5:00 in the morning. The foundation trench excavated that night must be restored in the morning, the road paved that night must be put into use in the morning, and as much material as needed must be taken out of the stock ground that night, and the surplus would have to be brought back to the stock ground and packed neatly before dawn. Strict and special construction arrangements let management side, construction side and design side often "meet late at night in the site, dawn here is buzzing with excitement".

西单商业区

Xidan Business District

因接触而诧异
——怎么这样

提起西单商业区，生活在北京或是来过北京的人都再熟悉不过了。这个一直作为北京商业形象代表的地区，因为2008年奥运会的原因，提前两年开始了一次较大规模的环境提升工程。我们有幸从城市设计开始，赢得了方案竞赛的第一名，从而与西单结下了缘分。

然而，当深入现场进行调研时，我们却惊诧地发现，代表首都商业形象的西单，实在是存在太多的问题。首先是交通问题，和王府井大街不同，西单的街道是城市主干道，无法改成步行商业街。现状人车混行、机非混行，存车设施短缺，交通隐患严重；其次是街区功能问题，市政设施老化，挤占人行空间，城市家具缺乏，道路凹凸不平；再次是开放空间问题，公共空间缺乏，休闲设施不足，文化意味缺失；最后是形象问题，建筑广告杂乱无章，门店牌匾各行其是，照明设施陈旧落后，标志标识不成体系，植物绿化差强人意……

Surprised by the Contact — why it is like this

The Xidan Business District is familiar to anyone who lives in or has been to Beijing. The district, which has long been a representative of Beijing's business image, began a large-scale environmental improvement project two years ahead of schedule because of the 2008 Olympics Games. We were lucky to start with urban design and won the first prize in the scheme competition, thus forming a relationship with Xidan.

However, when we went into the site and carried out the survey, we were surprised to find that Xidan, which represents the commercial image of the capital, had too many problems. First, there is the traffic problem. Unlike Wangfujing Street, the streets in Xidan are the main street of the city and cannot be turned into commercial pedestrian streets. The current situation is that pedestrian and vehicle traffic mixed, motor and non-motor vehicles mixed, car storage facilities supplied shortly, hidden traffic trouble is serious. The second problem is the function of the block, including the aging of municipal facilities, the pedestrian space crowded out, the lack of urban furniture, uneven roads. The third is the problem of open space, lack of public space, lack of leisure facilities and lack of cultural significance. Finally, there is the problem of image, the architectural advertising is disordered, the store plaques act in their own way, the lighting facilities are old and backward, the signs and markings are not systematic, the plant greening is unsatisfactory…

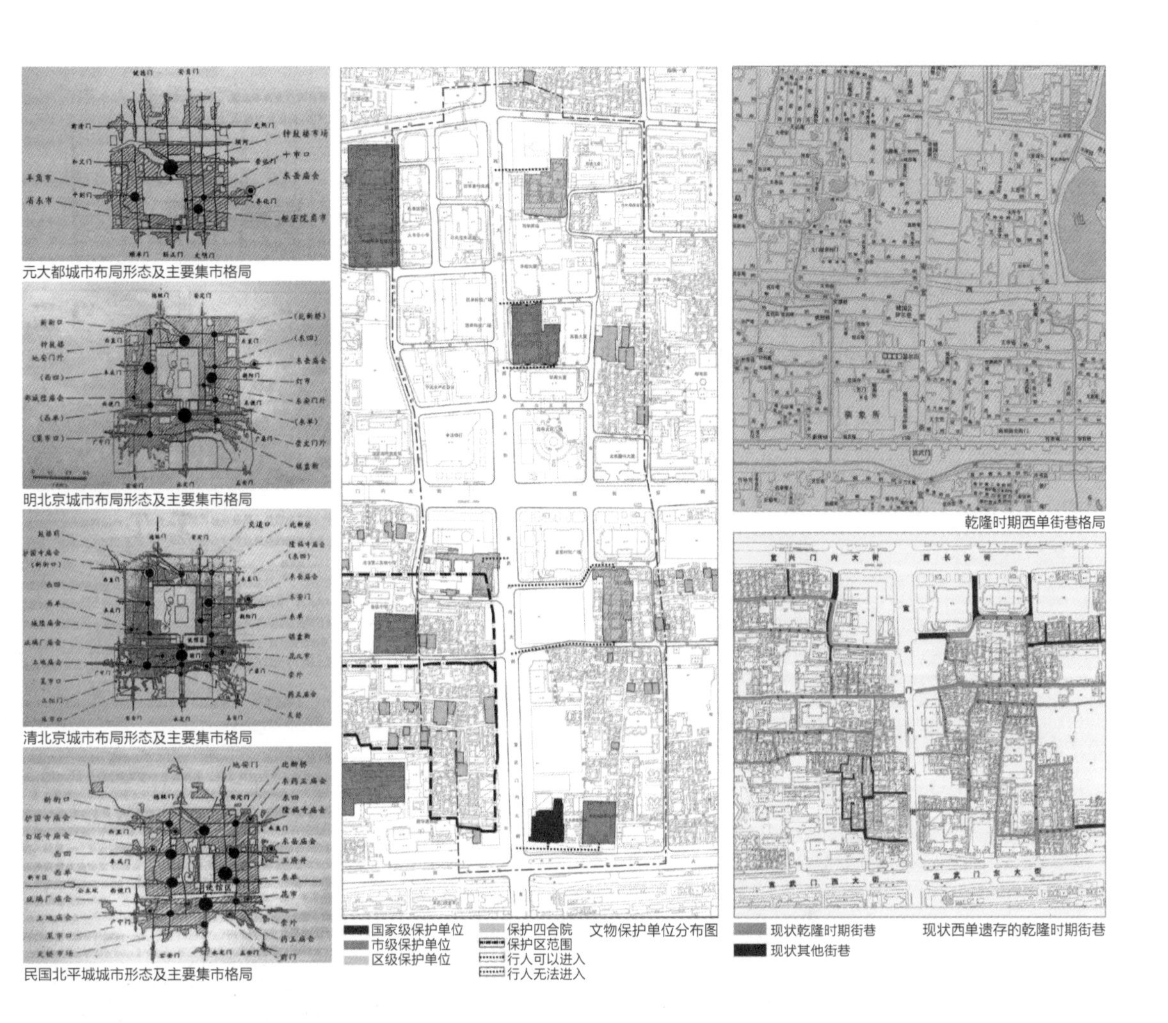

元大都城市布局形态及主要集市格局

明北京城市布局形态及主要集市格局

清北京城市布局形态及主要集市格局

民国北平城城市形态及主要集市格局

文物保护单位分布图

乾隆时期西单街巷格局

现状西单遗存的乾隆时期街巷

因研究而期待
——应该这样

在经过深入的分析和系统的城市设计后，我们发现西单商业区不可能单靠完成某一个单项设计就能实现整体品质的提升，它是一个系统问题，要用系统的方法去解决。于是我们提出“3C共融”的理念，即通过文化(Culture)、人本(Customer)、商业(Commerce)的融合来解决西单商业区所面临的各种问题，其中“文化”是动力，“商业”是基础，“顾客”是根本。

提出设计理念后，我们通过优化交通环境、完善市政设施、整治重要节点、美化街区景观4个方面的系统设计，不断将设计理念转化为可操作的设计方案。

（1）优化交通环境。此项工作是环境整治的重点，我们提出改造已有的过街天桥、增设新的天桥、增加室外自动扶梯等以人为本的措施完善了二层过街步行系统。进而在西单北大街增设中心隔离，彻底实现了人车分流，解决了多年人车混行的局面，为其他系统的设计提供了有效支撑。

（2）完善市政设施。在实现人车分流的前提下，我们又提出通过西单商业区主街变电箱的迁移，尽可能增加行人的通行空间；并通过管线入地等基础设施的改造与完善，为街区的进一步设计创造了必要的条件。

（3）整治重要节点。街道空间是线性的，线性空间中的节点就成了商业区活力的源泉。为此我们提出整治节点空间是提升街区品质的关键，并规划了4处节点空间：一是结合对国家级文物的保护和修缮，改造民族大世界；二是改造西单文化广场，进一步提升环境品质；三是新建南堂教堂外围广场，展现其独特的景观风格；四是拆除老四团小楼、电话局南侧小楼，增加街区开放空间。

（4）美化街区景观。在完成上述基础工作的前提下，我们从3个方面对街道空间展开了具体的设计：一是重新整治沿街建筑立面并规范广告牌匾的设计；二是通过铺装、绿化、城市家具、标识、雕塑等的系统设计使商业区形成富有文化内涵的独特个性；三是通过地面、橱窗、连廊、楼体4个层次的立体照明系统，提升街区夜间的商业活力。

Expecting by the Research — it should like this

After in-depth analysis and systematic urban design, we found that it is impossible for Xidan Business District to improve its overall quality by completing a single design. It is a system problem that needs to be solved in a systematic way. Therefore, we put forward the concept of "3C integration", that is, to solve various problems faced by Xidan Business District through the integration of Culture, Customer and Commerce, in which "Culture" is the impetus, "Commerce" is the basics and "Customer" is the foundation.

After putting forward the design concept, we continuously transformed the design concept into an operable design scheme through the systematic design of optimizing the traffic environment, improving municipal facilities, renovating important nodes and beautifying the street landscape.

Optimizing the traffic environment. This work is the key point of environmental remediation. We put forward the people oriented measures, such as renovating the existing overpass, increasing new overpass and outdoor escalator, to improve the two-floor pedestrian system. Furthermore, the center isolation was added in Xidan North Street, which completely realized the separation of people and vehicles, solved the situation that people and vehicles mixed for many years, and provided effective support for the design of other systems.

Improving municipal facilities. On the premise of realizing the separation of people and vehicles, we proposed that through the relocation of the main street substation in Xidan Business District, the space for pedestrians can be increased as much as possible. And through transformation and improvement like burying the pipeline into the ground and other infrastructure, the necessary conditions were created for the further design of the block.

Renovating important nodes. The street space is linear, and the nodes in the linear space become the source of the vitality of the business district. For this reason, we proposed that remediation node space is the key to improve the quality of the block, and planned four node spaces: first, the protection and repair of national-level cultural relics should be combined to transform the National World; second, the transformation of Xidan cultural square to further improve the quality of the environment; third, newly build the outer square of the south church, to show its unique landscape style; fourth, demolish the old fourth group building and the south building of the telephone office to increase the open space of the block.

Beautifying the street landscape. On the premise of completing the above basic work, we carried out specific design of street space from three aspects: first, renovate the facades of buildings along the street and standardize the design of advertising plaques; second, through the systematic design of paving, greening, urban furniture, signs and sculptures, make the commercial district form a unique personality with rich cultural connotation; the third is to enhance the commercial vitality of the block at night through the three-dimensional lighting system of the ground, window, corridor and building.

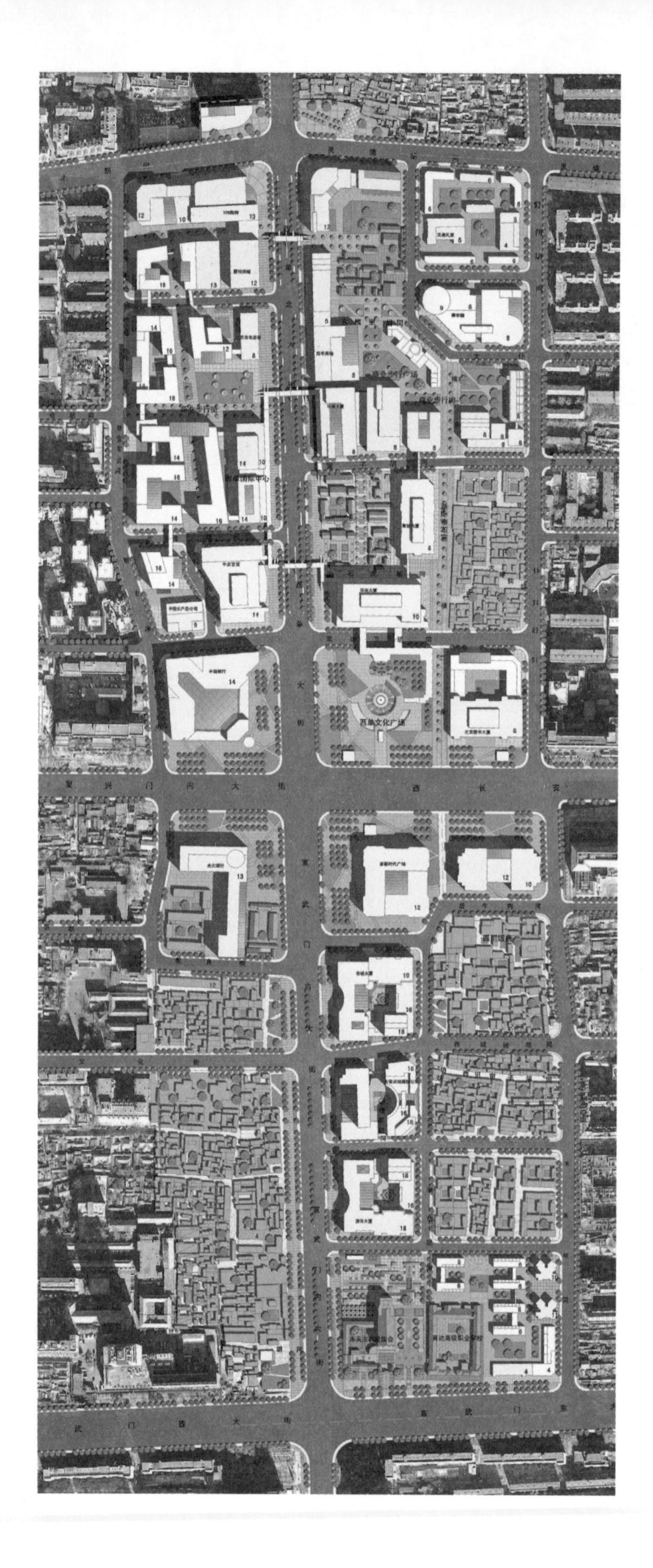

西单文化广场
复 兴 门 内 大 街
西 长 安

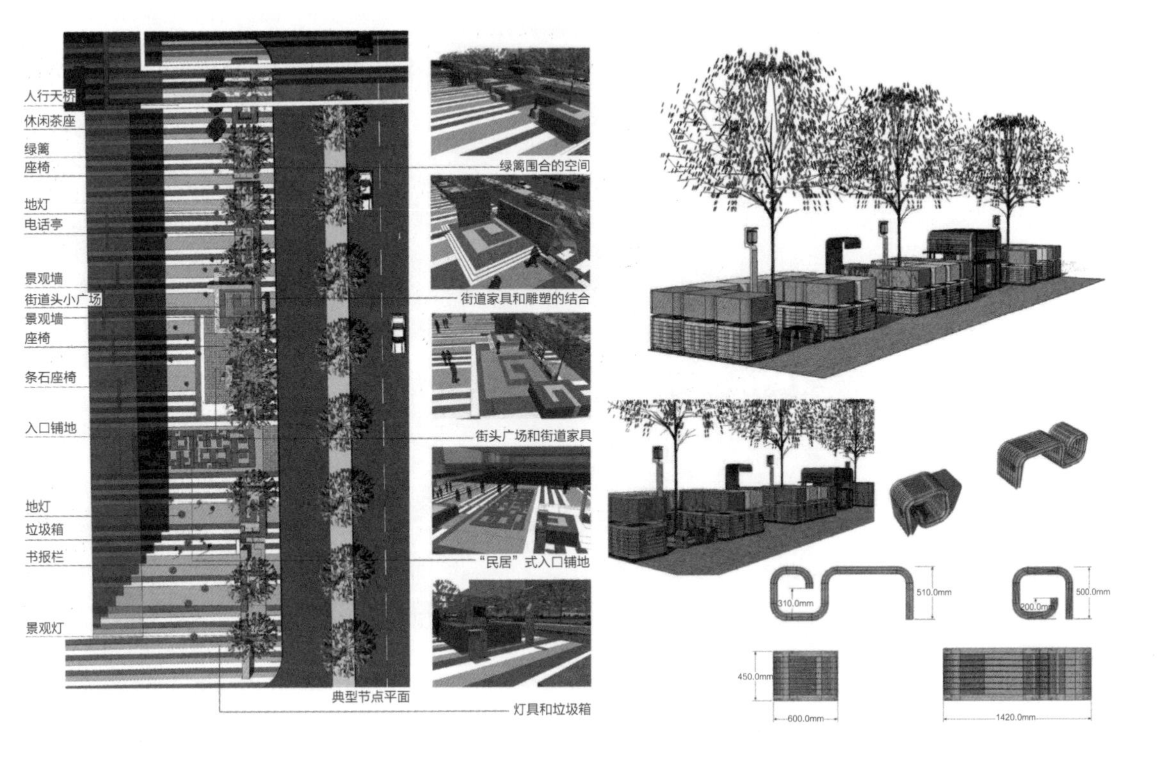

因实施而欣喜——总算这样

经过两年半的设计深化、修改以及无数次的汇报、交流、评审、现场施工配合，我们终于在奥运前完成了西单商业区环境整治的大部分工作，实现了人车分流和人性化的步行体系，开放了西单文化广场、南堂教堂等公共空间节点，街区景观面貌彻底改观。正是系统协调和功能为先的人本意识、历史保护和文化传承的设计理念、创新精神和艺术化的设计策略，使得西单商业区以崭新的面貌迎接奥运会的到来。

Rejoice in the Implementation — it likes this finally

After two and a half years of design deepening, modification and countless times of reports, exchanges, reviews, on-site construction cooperation, we finally completed most of the environmental renovation of Xidan Business District before the Olympic Games, realized the separation of people and vehicles and the humanized pedestrian system, opened Xidan Cultural Square, Nantang Church and other public space nodes, and completely changed the landscape of the block. It is the humanistic consciousness of system coordination and function first, the design concept of historical protection and cultural inheritance, the innovative spirit and artistic design strategy that make Xidan Business District welcome the Olympic Games with a new look.

如今的西单已经摆脱了昔日脏乱拥挤的形象，我们欣喜地看到：人行天桥的增加和室外自动扶梯的设置，人车终于互不干扰；步行空间中，购物步行区域、休闲绿化区域、无障碍通道、快速步行带等依次排布，井然有序；尤其是休闲绿化区域，将绿化、灯具、休闲座椅、垃圾箱、报刊亭、电话亭等街道设施巧妙地融为一体，结合西单LOGO特殊设计的座椅和树箱，展现出独特的时尚和文化；西单文化广场再现风采，南堂教堂外小广场上的镜面水池倒映出古老的穹顶，草坡和树阵下，人们享受其中；夜色降临，商业区楼体的广告标识与街面灯光交相辉映、绚丽多彩，我们的西单魅力非常……

Today, Xidan has shed its dirty and crowded image. We are glad to see: footbridges are increased and outdoor escalators are set up, people and vehicles finally do not interfere with each other; in the walking space, shopping walking area, leisure greening area, barrier free access, and fast walking belt are arranged in order; especially in the leisure greening area, the street facilities such as greening, lamps, leisure seats, dustbins, newsstands, telephone booths and so on are skillfully integrated into one, and the specially designed seats and tree boxes combined with Xidan LOGO are to show the unique fashion and culture. The Xidan Cultural Square reproduce the elegant demeanour, with the mirror pool on the square outside the Nantang Church reflecting the ancient dome, the grassy slope and the array of trees for people to enjoy; the night is coming, the advertising signs of the commercial district buildings and the street lights enhance each other's beauty, bright and colorful, our Xidan is very attractive···

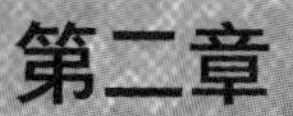

第二章

情感的外化

Chapter Two

Externalization of Emotion

万科公园五号售楼处小广场

Small Square near Sales Office No.5 of Vanke Park

“面儿”

我们的讨论起始于“面儿”，因为这个项目的核心问题就是既要满足城市的“面儿”，又要满足业主的“面儿”，当然还得满足自己的“面儿”。一直以来住宅项目售楼处的体验空间设计大多较为套路，大片的草坪和花卉、成组的大树和水池等，然而万科公园五号售楼处前却没有这样的空间，这就迫使我们另辟蹊径，以更纯粹、更自由的方式展开想象。

The "Face"

Our discussion began on the "face". Because the core problem of this project is to satisfy the "face" of the city, the "face" of the owner and, of course, the "face" of ourselves. The experience space design of the sales office of residential projects has been mostly routines, with large lawns and flowers, groups of trees and pools. However, there is no such space in front of the sales office No. 5 of Vanke Park, which forces us to explore a new way to develop our imagination in a purer and freer way.

“城市的面儿”：

“业主要面儿，但是这城市的面儿也太小了，才 800 平方米，能出什么效果。”这是接到任务后的第一反应。项目位于北京大望路的核心地带，面朝朝阳公园和豪宅棕榈泉展开的城市界面给设计提出了更高的要求。虽处高端地段但是城市界面杂乱不堪、交通混乱、噪声严重。如何给城市带来不一样的“面儿”成为我们的第一个思考点。

“业主的面儿”

“这是顶豪！北京地王！时尚策源地！”用高品质楼盘的售楼处景观来形容已经不能表达业主的需求了。

“除了展示楼盘与洽谈的功能要求，风格应配合建筑立面及室内设计，打造优雅从容的舒适室外空间，还得吸引眼球！过目不忘！最主要的是‘不一样’！得让客户来了‘有面儿’！”

“自己的面儿”

“其实也不难，就做个 artdeco 的就行了吧！就是超五星酒店那样儿的。没风险、有效果、好控制！”这是业主最终的决策之一！一个这么小的项目，其实可以这样做！但是当时我们觉得对不起“自己的面儿”。

"Face" of the City

"Owner needs face, but the face of this city is too small, just 800 square metre, what effect it can produce", this is the first response to the assignment. Located in the heartland of Dawang Road in Beijing, the project's urban interface facing Chaoyang Park and palatial palm springs demands a higher level of design. Although it is in the high-end area, the urban interface is very messy, traffic is chaotic and noise is serious. How to bring a different "face" to the city became our first thinking point.

"Face" of the Owner

"This is the best! The king of land in Beijing! Fashion source!" The description of sales office landscape of high-quality real estate can not express the needs of the owners.

"In addition to the functional requirements of displaying the building and negotiating, style needs to be fitted with the building facade and interior design. Not only create elegant, easy and comfortable outdoor space, but also attract attention of people! Impressive! The most important thing is 'different'. It need to let the owner has 'face' when coming."

"Face" of Ourselves

"In fact, it is not difficult to do an artdeco! It's like a super five-star hotel. No risk, effective, good control!" This is one of the final decisions of the owner! For such a small project, we can do like that in fact! But we felt sorry for our "face" at the time.

“块儿”

面对街角空间复杂的人流车流、噪声污染、没法迁走的电线杆子、满地的小广告，以及又要满足这、满足那的各方需求，我们展开了头脑风暴。首先，设计得“有劲儿”，也就是有体量感。其次，界面要错动，有冲击力、有力量感、有构成感和时尚感。第三，要利用艺术的手法，能吸引眼球，充满荷尔蒙和魅力。

由着上面的思考不断地推敲，我们将视线着眼于建筑立面的模块化处理。“干脆做‘块儿’吧”！“形态构成、模块堆叠、完整语言”，最终成为设计概念与母体。于是提取建筑立面合理的模数化尺寸，形成与建筑立面协调统一的立方体模块堆叠，通过与建筑夹角的变换增添了景观的灵活性，进而将此模数推而广之，地面铺装、树阵排布均采取此模数为单元进行组合，同时精于对空间的准确控制和景观元素单元的推敲组合，实现对场地最大化整合利用的思考而不是简单的美化装饰环境，使场地有趣、大气、合理、有机、统一，一气呵成！

Module

We brainstormed about the complexity of people and traffic flow on street corners, the noise pollution, the telegraph poles that couldn't be moved, the small advertisement all over the ground, and the need to meet this and that. First of all, the design needs to be "strong", that is, a sense of volume. Next, the interface needs to be stagger, having impulsive force, having sense of potency, having constituted feeling and fashionable feeling. The third, artistic technique is necessary, which can attract eyeball and have hormone and glamour.

Based on the above considerations, we focus on the modular treatment of the building facade. "Just do the module"! "Form composition, module stack, complete language" eventually became the design concept and matrix. Therefore, the reasonable modular dimensions of the building facade are extracted to form a stack of cube modules in harmony with the building facade. Through the transformation of the included angle with the building, the flexibility of the landscape is increased, and then the module is generalized. The floor paving and tree array arrangement are combined with this module as a unit, at the same time; it is good at precise control of space and careful combination of elements of landscape. The maximum integration and utilization of the site is playing and thinking, rather than simple environment beautification and decoration, making the site interesting, atmospheric, reasonable, organic, and unified at a time!

No.5 PARKFRONT 万科公园五號
銷售中心

“范儿”

“那马路口堆了一堆‘金块’”！

“那路口挺有趣的！”“很时尚！很艺术！”

“看起来很高端！格调挺高，这房子不便宜吧。”

建成的效果很快得到反馈，兴奋之余审视寻找起初的想法是否得到展现，答案是肯定的，模块化的组合使场地得到统一，与建筑获得一致。售楼处整体氛围与想象中一致，也获得了使用者的赞同。最重要的是它有了自己应该有的“范儿”！

Style

"There is a pile of gold nuggets at the crossing!"
"That corner is interesting!" "Fashionable! It is very artistic!"
"It looks high-end! It's very stylish. It's not cheap, is it?"
Feedback on the results of the construction was soon received, and the excitement was followed by a review to see if the original idea had been demonstrated. The answer is yes. The modular combination unified the site and made it consistent with the building. The overall atmosphere of the sales office is consistent with the imagination and has been approved by the users. The most important thing is that it has the "style" it should have!

郑州绿博会八一园

Bayi Garden in Zhengzhou Green Expo

寻迹

“八一园”是第二届郑州·中国绿化博览园的军区展园，位于绿博园东部展区范围内，用地呈三角形，面积为4179平方米。提到“八一”，我们就会想到中国人民解放军军旗和军徽的主要标志，充满神圣、庄严和纪念意味。于是我们以“绿色走进军营”为核心主题，以体现人民军队的精神特点为核心，表达新时期人民军队的综合实力。

园子所需承载的内容很丰富而用地又很紧凑，如何通过最具有感染力的形态来表现人民军队的精神面貌成为设计的首要难点；形态语言需要与军队硬朗的形象相匹配，需要通过视觉冲击力震撼人心，引起共鸣。

众多线索纷繁汇聚，寻找足迹成为共识，不忘初心追根溯源，我们选择了人民军队的使命特征作为主脉，即“对外保卫国家安全与领土完整，对内维护国家稳定与服务人民。”并据此将展园空间划分为对外展示界面——“筑界”和内部展示空间——“塑心”两部分。

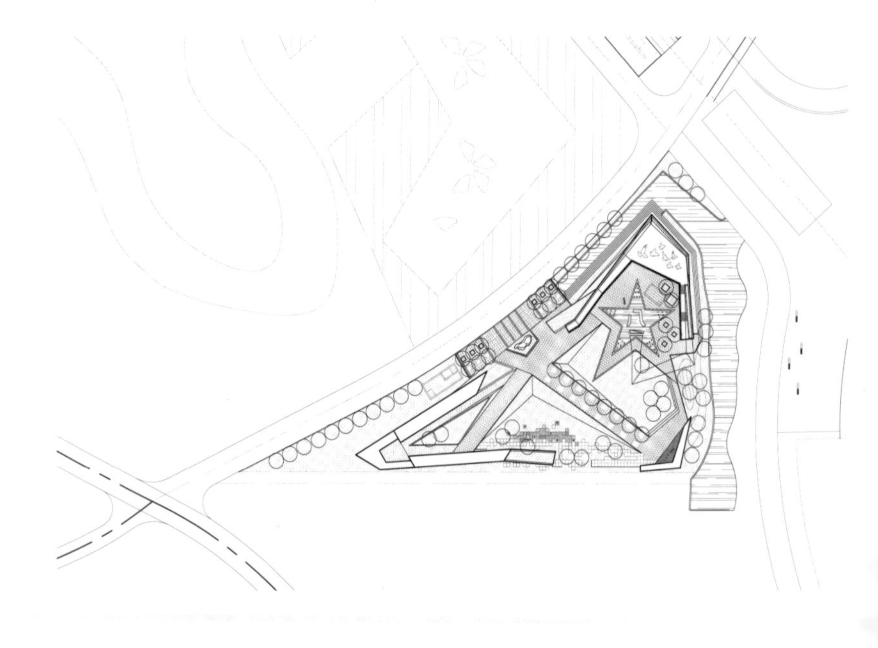

Looking for Footprints

"Bayi Garden" is the military exhibition garden of the Second Zhengzhou Green Expo of China, located in the eastern part of the exhibition area of the Green Expo, with a triangular area of 4179 square meters. "Bayi" reminds us of the main symbols of the flag and emblem of the people's liberation army, which are full of sacred, solemn and commemorative meanings. Therefore, we take "green into the barracks" as the core theme, and take embodying the spiritual characteristics of the people's army as the core, to express the comprehensive strength of the people's army in the new era.

The contents that the garden needs to express are rich but the site is compact. How to express the spiritual outlook of the people's army through the most appealing form has become the primary difficulty of the design; the form language needs to match with the strong image of the army, and needs visual impact to shock people and arouse sympathy.

Many clues are numerous and complex convergence, and looking for footprints become a consensus. Do not forget the original intention and trace the source. We chose the mission characteristics of the people's army as the main theme, that is, "to safeguard national security and territorial integrity abroad and to safeguarding national stability and serve people at home". Accordingly, the exhibition space is divided into two parts: the external display interface — "building boundary" and the internal display space — "shaping core".

筑界

对外展示界面采用钢结构加耐候钢板围合展园边界，通过硬朗、锋锐而又不失厚重的折线造型寓意人民军队保卫祖国、捍卫和平的使命，整体造型根据用地形态抽象而来，将战舰、战车、战机等形体特征融于其中，借鉴建筑造型手法，通过局部悬挑、折板、起翘，增加形态的变化，融入展示功能，增强视觉冲击力，象征人民军队刚强的一面。

Building Boundary

The external display interface is enclosed by steel structure and weathering steel plate. The strong, sharp and massive broken line shape implies the missions of the people's army that defend the motherland and defend the peace and the overall shape is abstract according to the land form, integrating the physical characteristics of battleship, chariot and fighter into it, drawing lessons from architectural modeling techniques. Through partial cantilever, folding plate and lifting, the change of shape is increased, and the display function is integrated to enhance the visual impact and symbolize the strong side of the people's army.

塑心

进入展园内部，核心为军徽广场，五角星形的镜面水池，倒映着和平主题雕塑，雕塑采用耐候钢板，整体造型简洁纯粹，浮雕和平鸽图案象征保家卫国的和平使命。广场周边是绿地展区，由三块主要绿地组成，作为“绿色走进军营”核心主题的集中展示，分别赋予绿色军营、荒漠绿洲、果岭飘香三个主题。绿色军营展区通过军队训练场景的模拟表现，结合林下绿荫空间布置形成展园的休憩空间，集雕塑、铺装、大乔木的组合，寓意军队刻苦训练、团结一致的日常军旅生活。荒漠绿洲展区借助五种铺装色彩寓意五色土地象征祖国广袤的疆域，内层为白沙，围绕绿地，寓意荒漠变绿洲，表达人民军队在新时期生态与绿化建设中的作用。果岭飘香展区运用大量乔木及色叶树打造丰富的植物景观，结合几何形抬高的地形，表达新时期军队荒山育林的绿化成就。

众多展区内设计有军旅纪念意义的小品和群雕，成为展园的活化激发点，使参观者步入园区犹如步入绿色军营，强调画面感和镜头感。借助几何绿地的边坡，结合高差设置记录墙，镌刻着不能忘却的记忆，形成内向型的线性空间，创造宁静的怀念与纪念氛围。

“筑界”“塑心”使项目获得相对完整的意义表达，作为情感外化的一种尝试，寻找形态表达需要遵循的内在逻辑，寻找精神内涵，继而外化于形体的构成与各个元素的并置关系，形成逻辑化的空间表达。

Shaping Core

Inside the exhibition garden, the core is the military emblem square, with a five-pointed star-shaped mirror pool, reflecting the peace-themed sculpture. The sculpture is made of weathering steel plate, whose overall shape is simple and pure, and engraving dove pattern symbolizes the peace mission of protecting the family and safeguarding the country. Around the square is the green space exhibition area, which is composed of three main green spaces. As a concentrated display of the core theme of "green into the barracks", three themes of the green barracks, desert oasis, and green fragrance are given respectively. Through the simulation of the military training scenario, the green barracks exhibition area forms the rest space of the exhibition garden in combination with the layout of the green shade space under the forest. The combination of sculpture, paving and large arbors symbolizes the daily military life of the army with hard training and unity. The exhibition area of desert oasis symbolizes the vast territory of the motherland with the aid of five kinds of paving color implication of the five colors of the land. The inner layer is white sand and the surrounding green space symbolizes the transformation from desert to oasis, expressing the role of the people's army in the construction of ecology and greening in the new era. A large number of arbors and color-leaved trees are used in the green fragrance exhibition area to create a rich landscape of plants. Combined with the geometric elevation of the terrain, it expresses the greening achievements of the army's afforesting in barren mountains in the new period.

Many exhibition areas are designed with military memorial landscape decoration and group sculptures, which become the activation and stimulation points of the exhibition park, making visitors walk into the park as if they were walking into the green barracks, emphasizing the sense of picture and the sense of shot. With the help of the slope of the geometric green space, a record wall is set by the height difference, engraving with unforgettable memories, forming an introverted linear space and creating a tranquil atmosphere of yearning and remembrance.

Building boundary and shaping core make the project obtain a relatively complete meaning expression. As an attempt to externalize emotions, it should look for the inner logic that form expression needs to follow, look for the spiritual connotation, and then externalize the structure of the form and the juxtaposition of each element to form a logical spatial expression.

光大中心
Everbright Centre

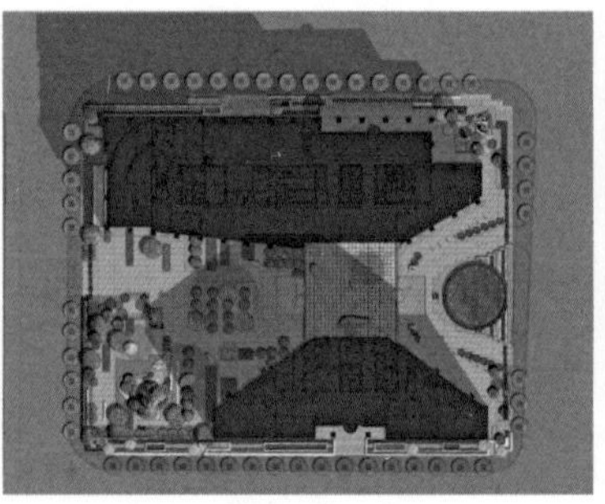

心中的“光大”

从1992年开始，众多知名金融企业逐渐聚集到北京西二环边，慢慢形成了著名的金融街。2009年，金融街东北角建了个端正的大楼，名叫“中国光大银行中心”，作为景观设计师，心中的“光大”不只是银行名称，还有对金融的理解以及对环境塑造的把握。

"Everbright" in Heart

Since 1992, many well-known financial enterprises have gradually gathered at the edge of Beijing's west second ring road, forming a famous financial street little by little. In 2009, an upright building called "China Everbright Bank Center" was built at the northeast corner of financial street. As a landscape designer, the "everbright" in heart is not only the name of the bank , but also the understanding of finance and the grasp of environmental shaping.

聊出的契合

为了精准理解“光大”的精髓，我们和光大银行行长聊，和SOM的建筑师聊，和日建的室内设计师聊，和之前做过这个地块的景观设计师聊……终于，聊出了契合，这种契合使得后面的景观设计犹如行云流水般应运而生。首先是建筑的解读：外观方正端庄、内向弓背相对的对弓形A、B双塔建筑对场地具有绝对的控制力。室内设计师山梨先生的原则是延续这样的张力，在内部空间柔和地体现出中国文化的含蓄；老行长对中国书法的挚爱也寄情于光大门前那20余吨的厚重石壁上。融合所有相关者的理念，我们逐渐达成共识，追求场地自身散发出的张弛之势，追求银行企业应有的气宇轩昂，追求内外融合的文化气韵。

“地方”与“天圆”

达成的共识要转化为景观语言，我们尝试将场地作为绿色的基底，其上托起一个能承托寄寓的重器之型，用以表达对光大银行“EB”（EVER BRIGHT）精神，即“共享阳光，创新生活”的解读。最终，我们选择了“天圆地方”的形式。“地方”是基底，“天圆”是圆池。圆池暗合山梨先生之含蓄，用水于无形，表情丰富。尽可能扩展，宛若一个端起的金盘，每日“光大”，迎向朝阳。

Fit after Talking

In order to accurately understand the essence of "everbright", we talked with the President of Everbright Bank, the architects of SOM, the interior designers of NIKKEN, and the landscape designers who had worked on the site before… Finally, we found a fit, which makes the landscape design after that rise at the right moment like flowing water. First is the interpretation of architecture: the appearance is square and dignified, and the inner double towers A and B with opposite arched backs have absolute control over the site. Mr. Yamanashi's principle in interior design is to continue such tension, reflecting the implicit Chinese culture in the soft interior space; the old President's devotion for Chinese calligraphy also lies in more than 20 tons heavy stone wall in front of the gate of Everbright Bank. By integrating the ideas of all stakeholders, we gradually reached a consensus to pursue the relaxation of the site itself, the required straight and impressive looking of the banking enterprise, and the cultural charm of internal and external integration.

Orbicular Sky and Rectangular Site

The consensus reached need to be translated into landscape language. We try to use the site as the green base, on which a shape of heavy container can be supported to express "EB" ("EVER BRIGHT") spirit of Everbright Bank, that is, the interpretation of "sharing sunshine and innovating life". In the end, we chose "orbicular sky and rectangular site". Rectangular site is the base, and orbicular sky is round pool. Round pool suits the implicit of Mr. Yamanashi, imperceptibly utilizing water with rich expression. Expand as far as possible, like a held-up gold plate, daily "ever bright", to meet the sunrise.

王府壹号
Wangfu No.1

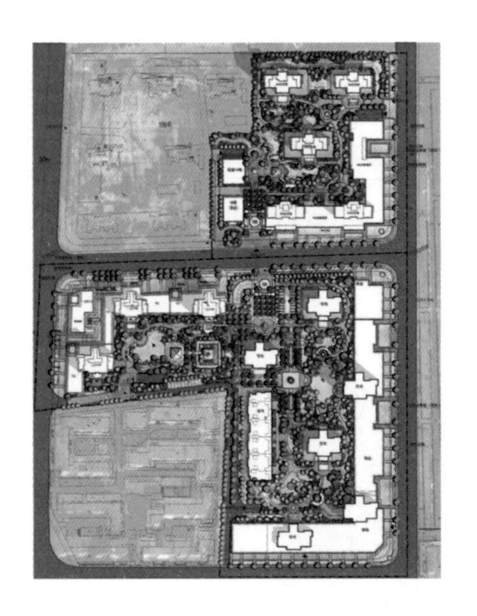

王府的难

融创王府壹号项目的设计任务始于2011年，放眼回望那时候的中国地产行业正处于由“黄金时期”向“白银时期”转变的关键节点上。在那之前的地产景观设计还是有比较大的创作空间，无论是我们设计万科紫台的国学气韵还是魅力之城的大院文化，抑或洋浦花园的北欧情趣、公园五号的时尚创新等等，都呈现更为多元的开放视野与创作角度。而大概是从这个时期开始，中国地产景观逐渐“拘谨起来”，也慢慢地走向如今的“千篇一律”。

项目位于天津老城区核心地带，由于比邻庄王府而得名“王府壹号”，通过这高大上的案名也能窥探开发者对它的定位与期许。当然，“王府”有“王府的难”，作为当年天津的地王，定位是中国北方的住宅标杆，也是融创壹号院品牌的始点之一，在当年，如何打造豪宅，清楚的人并不多。可能也是由于这么高的定位，使得从业主到之前的设计师都有点“过于紧张”。法式的售楼处、“星河湾式”的前区、略带中式韵味的室内设计以及“东南亚风格”的景观示范区，多种元素叠加带给我们的初始印象是“有点乱”！

Difficulties of Wangfu

The design task of Rongchuang Wangfu No.1 began in 2011. Reviewing back, China's real estate industry was at a critical juncture in its transformation from a "golden age" to a "silver age" at that time. Before that, the real estate landscape design still had a relatively large creative space. Whether the traditional Chinese artistic conception of Vanke Zitai we designed or the courtyard culture of the charming city, or the Nordic appeal of Yangpu Garden and the fashion innovation of Park No. 5, we presented a more diversified open vision and creative perspective. And from this period, Chinese real estate landscape gradually becomes "overcautious", and slowly goes toward today's "cookie-cutter".

The project is located in the core area of the old city of Tianjin, and is named "Wangfu No. 1" because of the neighboring Zhuangwangfu. Through this lofty project name, we can also see the developer's positioning and expectations for it. Of course, "Wangfu" has its difficulties. As the prime site of Tianjin in those days, its positioning is the residential benchmark in north of China and also one of the starting points of Rongchuang No.1 Courtyard brand. In those years, not many people know how to make a luxury real estate. Perhaps because of such a high positioning, from the owner to the previous designers are a little "too nervous." The French style sales office, "star river bay style" front area, slightly Chinese style interior design and "southeast Asia style" landscape demonstration area, a variety of elements to give us the initial impression is "a little messy"!

以静制动

项目虽地理位置优越，但周边是嘈扰、杂乱而富有生趣的市井之像，初到之时会给人一种一头扎进历史场景中的感觉。面对项目风格定位的“动态摇摆”与周边多元的环境因素，我们该如何“一招制胜”，在满足各方需求的同时营造出一定的叙事性与崇高感呢?

我们决定“以静制动”，用景观的手法营造传统王府空间的序列性，在空间非常有限的核心园林区结合下沉会所空间营造超尺度的平静水面，通过水面的尺度与空间的对比给人“万物静雅”的空间体验。在创作的过程中我们始终认为，要以一个“超·大”的“动作”来统领所有问题与矛盾，而这个动作的载体一定是“万事不争”之“水”，她要包容一切，要消解一切！一方面，这个水一定是“超静”的！一定是褪去铅华、温润剔透，像植入旧城的一块玉一样“镇住”所有不确定的“风格”与不确定的欲望，同时她又如一面反射万物的镜子，将所有的物体吸收进来，统一起来。当时，我们明确要用“去风格化”来解决和统领“风格之争”。另一方面，就是“大”，在这么小的空间做这么大的体量是有很大挑战的，但只有通过“对比”才能营造出一定的“戏剧性”，与外围杂乱的小街小巷形成超乎寻常的对比。

项目建成后，带团队回访之时，新入职的杨工惊呼“你们是怎么说服甲方的，真不容易！”但是，谁又知道甲方当时也是想“越大气越好，简洁才是高端”。事实证明，彼时彼刻的决策具有一定的试验性，但是我们都赢了！

Coping with All Motions by Remaining Motionless

Despite its good location, the project is still surrounded by a noisy, cluttered and interesting marketplace, which, when people first arrive, feels like a headlong plunge into a historical scene. Facing the "dynamic swing" of the project style positioning and the surrounding multiple environmental factors, how can we "win with one move" to meet the needs of all parties while creating a certain narrative and noble sense?

We decided to "cope with all motions by remaining motionless". Used the landscape technique to create the sequence of the traditional royal house space, but combined with the sinking club space in the very limited space of the core garden area to create a super-scale calm water surface. Through the contrast between the scale of the water surface and the space, give an "all things quiet and elegant" space experience. In the process of creation, we always believe that all problems and contradictions should be dominated by a "super big" "movement", and the carrier of this movement must be the "water" of "everything is indisputable", she must contain everything and dispel everything! On one hand, the water must be "very quiet"! It must be to return to innocence, to be warm and crystal clear, to "suppress" all uncertain "styles" and uncertain desires like a piece of jade planted in the old city; it is also like a mirror reflecting everything, absorbing all the objects and uniting them. At that time, we clearly want to use "de-stylization" to solve and lead the "battle of styles." On the other hand, it is "large". It is very challenging to make such a large volume in such a small space, but only through "contrast" can create a certain "drama", which forms an extraordinary contrast with the small and disorderly streets in the periphery.

After the completion of the project, when we led the team to pay a return visit, the new employee, Mr. Yang, exclaimed, "How did you persuade party A? It's quite hard!" However, who knew that party A also wanted "the nobler the better, simplicity is the high-end"? As it turned out, the decision making is experimental in some way at that time, but we all won!

航天城飞天园

Flying Garden in Aerospace Town

场地制约

航天城飞天园整体布局规则方正，建筑体量人，绿地分布零碎，植物品种单一，缺少大树；由于建设时间长，缺少原始资料，很多现场资料无法取得，尤其是地下管线的资料更是无从得知；而场地景观改造成功的关键就是如何收集和利用场地的雨水，并打造成特色的水景效果。

关键技术

航天城飞天园是在平地中打造水景，重点要解决水源、水形、水质净化、水循环等关键技术问题。通过软件模拟计算出一年的地表径流量、蒸发量以及现有供水水源的富余量来核定水面大小，由此保证水源充足。水的形态既要考虑地形的塑造又要反映航天的内涵，飞天水形正好满足这双重要求。大小水循环设施是保证良好水质的关键，但在此还需要结合水质净化处理装置以及大量的水生植物等多重技术，方能保证大雨过后半天即清的要求。

Site Constraints

The overall layout of the Flying Garden in Aerospace Town is square, with a large building volume and fragmented green space. The plant species is single and lacks large trees. Due to the long construction time and the lack of original data, a lot of site data cannot be obtained; especially the underground pipeline data is unknown. The key to the success of the site landscape transformation is how to collect and use the rainwater of the site, and create a characteristic waterscape effect.

Key Technology

Flying Garden in Aerospace Town is to create a waterscape in the ground, and the focus are to solve the water source, water form, water purification, water circulation and other key technical problems. Software simulation is used to calculate the annual surface runoff, evaporation and the surplus of existing water supply to determine the size of the water surface, so as to ensure adequate water supply. The form of water should not only consider the shape of terrain but also reflect the connotation of spaceflight. The shape of flying sky and water just meets this dual requirement. Large and small water circulation facilities are the key to ensure good water quality, but this also requires the combination of water purification equipment and a large number of aquatic plants and other multiple technologies, in order to ensure the requirement that the heavy rain will be clear after half a day.

自然表情

组合式关键技术利用后，航天城飞天园的水景呈现了高品质的自然表情。大循环和小循环分别保证了不同水位和不同气氛的要求，使得源头成为涌泉水景，过程净化形成自然溪流和生态叠水，溪流汇集到末端自然形成安静的湖面，通过缓坡绿地入水驳岸、阶梯式荷花种植台、水生植物种植区等手法，还原了水面的自然表情。

Natural Expression

After the use of combined key technology, the waterscape of Flying Garden in Aerospace Town presents a high-quality natural expression. Large and small circulation guarantee respectively the requirements of different water levels and different atmospheres, making the source become the water scene of spring, and purification process forms the natural stream and ecological overlapping water. The stream gathers to the end of the lake and naturally forms a quiet lake. By means of gently sloping green land into water revetment, stepped lotus planting platform and aquatic plant planting area, the natural expression of the water is restored.

重庆西永广场

Xiyong Square in Chongqing

山水重庆

西永中央广场的建设应该是重庆曾经如火如荼的“广场运动”的最后一个。在西永新区的城市中央建设大尺度城市公共空间，借以提速城市建设进程，带动周边土地增值。重庆因山地城市特征及山水格局闻名于世，因此在广场几个核心景点的创作上，我们自然而然地想到了重庆的山水，希望运用抽象的设计语言再现山城所独有的山水意境。其中位于广场北端起点的“奔涌之水”与位于广场南端的“五彩之池”最具特点，通过营造不同的水景表情塑造川渝地区独特的韵味。

Chongqing's Beauty in Mountain and Water

The construction of Xiyong Central Square should be the last of the "square movement" that has been in full swing in Chongqing. The construction of large-scale urban public space in the center of Xiyongxin district can speed up the urban construction process and promote the appreciation of surrounding land. Chongqing is famous for its mountainous city features and the pattern of mountain and water. Therefore, in the creation of several core scenic spots in the square design, we naturally think of the mountains and water of Chongqing and hope to reproduce the unique landscape artistic conception of the mountain city with abstract design language. Among them, the "surging water" at the beginning of the north edge of the square and the "colorful pool" at the south edge of the square have the most characteristics, through creating different waterscape expressions to shape the unique flavor of Sichuan and Chongqing.

奔涌之水

水具有无数的表情，或平静，或悠远，或灵动，或神秘。我们初次驻足场地之时，环顾四周群山，远眺嘉陵江，便有了将大山大水之美揽入怀中的冲动。于是在广场北侧水景的创作中，我们便利用场地原有的8米高差，营造水的壮阔之势，水由涓涓细流几经转折流淌后奔涌而下，拍打“山川”泛起层层浪花，其景象之壮阔会使观者不由想起长江之水的雄浑，想起嘉陵之水的悠远。

五彩之池

广场南侧的场地是以餐饮服务为主的休闲空间，广场中心的水景设计其实也是业主的命题作文，起初我们提出以“山形”装置为核心的提案始终未能通过，后期提出的几个水景形式也被认为过于普通，缺乏独特性。后来，我们将创作焦点聚集在地域性上，这个水景需要表情灵动、具有吸引力，还需要一种晶莹剔透的气质，由此我们想到了四川黄龙的五彩池！水池形体的穿插堆叠、水在五层水池之间的翻转流动、夜景灯光营造的五彩变化，不言自喻的文化认同，一切显得合理而自然，方案终于全票通过。

Surging Water

Water has countless expressions, either calm, or distant, or flexible, or mysterious. When we first stopped the site, looking around the mountains, overlooking the Jialing River, there would be the impulse to embrace the beauty of mountains and waters into arms. So in the creation of the water scene in north of square, we used the original height of 8 meters, creating the magnificent momentum of water. The water flows from trickles, turns several times, and surges down finally, slapping layers of waves on the "mountains". The scene of the magnificent momentum will make the audience couldn't help thinking of the powerful Yangtze River, thinking of the distant of Jialing River.

Colorful Pool

The site on the south side of the square is a leisure space dominated by catering services. The waterscape design in the center of the square is also the owner's proposition composition. At first, our proposal centered on the "mountain shape" device was never approved, and several waterscape forms proposed in the later period were also considered too common and lacking in uniqueness. Later, we focalize to the creative focus on the localization; the water scene not only needs a flexible and attractive expression, but also needs a kind of crystal clear temperament. From this, we thought of the Five Color Pool in Sichuan Huanglong! The interspersing and stacking of the pool, the overturn and flow of water between the five-layer pools, the colorful changes of night view lighting, and the self-explanatory cultural identity, all seemed reasonable and natural, and the scheme was finally unanimously approved.

拉萨饭店

Lhasa Hotel

初始印象

2009年的初秋，我们接到拉萨饭店改造的任务，由于布达拉宫周边与宗角禄康公园以及拉萨火车站等工程，对拉萨已经很熟悉了。到达拉萨饭店后第一印象是楼很实、院很小、树很大，需要改造的余地不多。院东侧三排杨树叶已经金黄，与西侧两棵百年老柳树遥相对视，在微风吹拂下发出悦耳的声音，这是场地上最美的景观，保护好这些树就是设计的初始印象。

朴素逻辑

为了完整地保护好现状树木，整体空间营造采用了分层施策的朴素逻辑。通过划定外围绿化保护带将现状绝大多数大树完好地保护起来，划定中间景观过渡带将大树和建筑协调起来，划定建筑入口重点区使入口景观灵动起来，划定主题庭院区使院落生动起来。如此朴素的设计逻辑将现场一切可以调动的景观要素充分调动起来，激活了每一个空间角落，使原本尺度不大、分布零散的各类空间有机串联形成一个整体，以期形成小中见大的效果。

Initial Impression

In the early autumn of 2009, we were given the task of renovating Lhasa Hotel. Lhasa has been very familiar with the Potala Palace, the Zongjiao Lukang Park and the Lhasa Railway Station. The first impression after arriving at Lhasa Hotel was that the building is very solid, the courtyard is very small and the trees are very big, so there is not much room for transformation. The leaves of three rows of poplars on the east side of the courtyard were already golden, and were looking at the two century-old willows on the west side in the distance, making a pleasant sound under the breeze. This was the most beautiful landscape on the site, and protecting these trees was the initial impression of the design.

Simple Logic

In order to protect the existing trees completely, the whole space is built with the simple logic of layering and targeted solution. Delimit the peripheral green protection zone to well protect the majority of the existing trees, delimit the intermediate landscape transition zone to harmonize the trees and buildings, delimit the key area at the entrance of the building to make the entrance landscape lively, and delimit the theme courtyard area to make the courtyard lively. Such simple design logic will fully mobilize all the landscape elements that can be mobilized on the site, activating every corner of the space, making the original small scale and scattered kinds of space organically form a whole, in order to form a small to see the big effect.

水石自然

空间的整体联系有了，细节上如何吸引入并体现当地特色，这是小尺度庭院设计的难点。由于之前的项目对拉萨进行过比较系统的调研，相比较而言，拉萨的“水与石”是有特点的景观元素，也是拉萨自然的朴素反映，如果适当地运用在庭院中，能够起到烘托情感、营造意境的作用。于是将六个大小不一的庭院围绕“水与石”赋予不同的主题，充分利用拉萨河中的自然荒石，或卧，或立，或三五成群，或主次对望，或前后错落，或高低起伏，多重组合手法打造“西藏山水”“哈达石林”“墨荷小苑”等不同的庭院意境，将初始的水石自然印象转化为景观情感，外化于庭院的空间表情中。

Natural Water and Stone

The space has got its overall connection, but how to attract people and reflect local characteristics in detail is the difficulty of small-scale courtyard design. Due to the systematic research on Lhasa in previous projects, comparatively speaking, Lhasa's "water and stone" is a characteristic landscape element and a simple reflection of Lhasa's nature. If properly used in the courtyard, it can play a role of causing emotions and creating artistic conception. So 6 courtyards of different sizes were given different themes around "water and stone", making full use of natural stones in Lhasa river, or lying, or standing, or groups, or looking at primary and secondary, or strewn at random before and after, or ups and downs. Multiple combination technique makes different garden artistic conceptions such as the "Tibet's mountain and water", "Hada stone forest", "small garden of lotus". The initial impression of natural water and stone is transformed into landscape emotion and externalized into the spatial expression of the courtyard.

航天城望山园
Wangshan Park in Aerospace Town

场地设问

从入口进入航天城，穿过修剪的低矮绿篱不经意间就会直接进入办公楼，楼前空旷的广场很难给人留下印象。但广场西南角“L”形下沉广场却引起了我们的思考：为什么是“L”形？与主楼立面存在视线对位关系？为什么仅下沉这点空间，功能是什么？为什么不种树？这种由现状肌理不由自主的思考成了设计构思的开始。

树石作答

要解决好广场上的问题，首先要从空间视线分析入手。从西门而入，将流线直接引入主办公楼，这是人流高效引导的做法，也符合员工固有的行为习惯，但是从视线上分析，这个流线的右侧配楼打破了左右广场大小不一的空间平衡，这里似乎需要一个竖向体量的元素来矫正平面视觉上的不平衡。于是，树阵和绿篱变成了不错的选择，用树阵来打造一个视觉轴线，使得西侧的人流顺着树阵自然而然地进入办公楼。从办公楼出来，左侧配楼和主楼形成一个“L”形建筑实体，而右侧却是空旷的“L”形下沉广场，透过广场直接看到远处的停车场，景观效果空而差，如何遮挡这一空而差的视线，并与建筑呼应，形成办公楼前的广场空间围合？假山置石成了必然选择。由此，树与石是解决广场问题的关键元素。

Questions About the Site

From the entrance entering into Aerospace Town, through the low cut hedge will inadvertently directly into the office building, the empty square in front of the building is difficult to leave an impression for people. However, the "L" shaped sunken square at the southwest corner of the square caused us to think: Why it is the "L" shape? Is there a para-position relationship of sight with the facade of the main building? Why only sink this area of the square? What is the function? Why not plant trees? This kind of spontaneous thinking from the existing texture became the beginning of the design idea.

Answers by the Trees and Stones

To solve the problems of the square, we should start with the analysis of spatial line of sight. Entering from the west gate, direct the streamline into the main office building, this is not only an efficient way to guide the flow of people, but also in line with the inherent behavior of employees. However, from the perspective of line of sight, the building on the right wing of this streamline broke the spatial balance of the left and right squares with different sizes, which seemed to require a vertical volume element to correct the imbalance in plane vision. Trees and hedges became good choices accordingly. Use the tree array to create a visual axis so that the flow of people in the west side naturally went along the tree array into the office building. Out of the office building, the main building and the building on the left wing formed an "L" shaped architectural entity, while the right wing is an empty "L" shaped sunken square. Looking directly through the square to the parking lot in the distance, the landscape effect is empty and poor. How to block out this empty and poor line of sight, and echo with the building, forming the square space in front of the office building? Rockery stones have become an inevitable choice. Therefore, trees and stones are the key elements to solve the square problem.

营林观景

设计策略确定后，为了保证良好的视线，46棵大银杏的移植成为关键。为实现全冠移植的目标，依据银杏的长势和丰满程度，采用“轻截枝、重疏叶”的移栽技术。苗木除伤病枝短截外，其他枝条一律轻截，甚至不截，枝叶依长势选择性疏剪1/2~2/3，以减少蒸腾作用，每棵树配备营养液和水位观测口，即时监测银杏的含水情况。经过综合养护技术的运用，最终保证了银杏的全部成活，形成了广场景观的绿色空间界面。

筑石望山

顺着银杏林向西南角看去，“L”形下沉广场正好与建筑相对，因形就势设计“L”形假山叠水，为建筑主要视线打造了“活”的对景，并很好地利用了原有肌理。假山设计强调“远观有势、近看有形、进入有景”的特色原则，通过采用大小不一的房山石层层叠砌，既考虑前后错落、高低起伏，又强调虚实对比、主次对望，既有盘山道、亲水台，又有石中树、石中水，并结合木质亲水平台、水生植物、造型油松等元素，打造了层次丰富、自然渗透的山石景观，使场地获得了新生。

Making Forest and Enjoying View

After determining the design strategy, in order to ensure a good line of sight, the transplantation of 46 large ginkgo trees became the key point. In order to achieve the goal of full crown transplantation, according to the growth and fullness of ginkgo, the technique of "weaken cutting branches, emphasize trimming leaves" was adopted. Except for the weak cut of diseased branches, other branches all weaken cut, or even not cut. Branches and leaves are selective cut 1/2-2/3 according to the growth, in order to reduce transpiration. Each tree is equipped with nutrient solution and water level observation mouth, real-time monitoring of the water content of ginkgo. Through the use of comprehensive technology, finally ensure the survival of all ginkgoes, and form the green space interface of square landscape.

Constructing Stones and Viewing Mountains

Looking from the southwest corner of ginkgo grove, "L" sunken square is just opposite to the building. According to the "L" shape and terrain, design "L" shape rockery and stacked water, this created a “live” opposite scene for the main view of the building, and well used the original texture. The design of rockery emphasizes the characteristic principle of "magnificent far away, tangible near, scenic getting into". By means of stacking rock of different sizes layer upon layer, considering strewn at random forward and backward and the ups and downs, and emphasizing the contrast of virtual and real, the sight line of primary and secondary. There are not only hillside roads and waterside platforms, but also trees near stones and water near stones, and the combination of wood waterside platform, aquatic plants, designed shape Chinese pine and other elements creates a rich, natural infiltration of the stone landscape, so that the site has a new life.

雁栖岛

Yanqi Island

礼石

从对自然山水的观照赞美，到人造山水，似乎是一种合乎逻辑的演变。人们感叹于自然山水的雄美壮阔，自然山石的层叠林立，想要迁移自然山水于庭院之中，才有了叠山理水的人造园林。

雁栖湖的叠石也是如此，因为雁栖岛经常有大型的国事活动，因而在此叠石，我们更关注“礼石”。“礼石”的开始，起于“用心”，既不屈服于自然之下，又不凌驾于自然之上。我们以此为原则，精选了泰山石、黄蜡石、斧劈石，经过现场的搭配摆放，才最终形成了收放有致的叠石效果。

Ritual Stone

It seems to be a logical evolution from the contemplation and admiration of the natural landscape to the man-made landscape. People are amazed at the magnificent beauty of natural mountains and rivers and the natural mountains and stones stacked in forest, and want to move into the natural mountains and rivers and courtyard. As a result, man-made gardens with stacked mountains and rivers are designed and built.

The same is true of the stacked rocks of Yanqi Lake. Because Yanqi Island often has large-scale state activities, so here, we pay more attention to the "ritual stone". The "ritual stone" starts from the "heart", which neither surrenders to nature nor surpasses nature. We took this as the principle, selected the Taishan stone, the chrismatite and the Fupi stone, through the collocation and placement at the site, and finally formed the perfect stone effect.

——泰山石的沉稳、凝重、浑厚，搭配造型松的千姿百态，别有一番韵味，叠石理水、松石搭配，是节点造景的不二之选。

——斧劈石则略显工整，高矮大小搭配极具设计感，与酒店的风格遥相呼应，尽显石头的质感。在人造的环境中，找寻自然的韵味。

——房山石形态各异、造型奇特、如圆似方，具备了“漏、透、瘦、皱”等传统赏石要素，是水岸、旱溪叠石的优质素材，石缝间搭配各色植被，与石头相映成趣，营造了自然之石野趣盎然的意境。

The combination of the composed, dignified, vigorous Taishan stone and the designed shape Chinese pine in different poses and with different expressions has a lasting appeal. The stacked stone and waterscape, the collocation of Chinese pine and stones, are the most wonderful choice of node landscape.

The Fupi stone is slightly neat; height and size are with a sense of design, and in line with the hotel's style, showing perfectly the texture of stone. Find the charm of nature in the artificial environment.

The Fangshan stone that has different forms, unique shapes, both round and square, with "leakage, penetration, thin, wrinkle" and other traditional stone elements, is the high quality material for the water bank, the dry stream, and stacked stone. Match between stone crevice of various kinds of vegetation, formed a delightful contrast with stones, and built the artistic conception of natural stone, wild and full of interest.

牙买加中国园林

Classical Chinese Garden in Jamaica

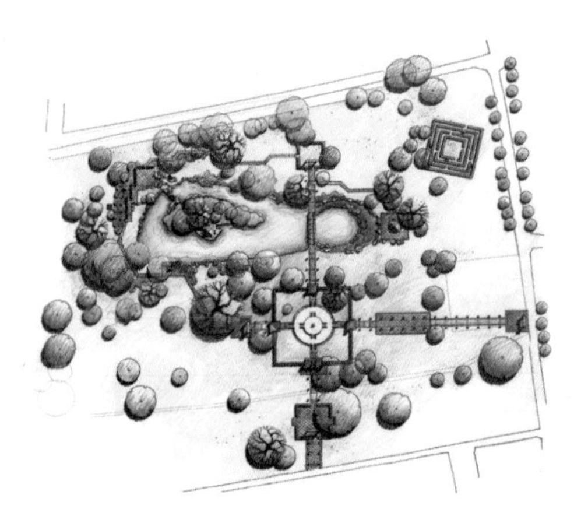

美丽的加勒比

2008年12月，我们从寒冷的北京飞抵加勒比海美丽的岛国牙买加，扑面而来的是温润的带着花香的气息，葱郁的热带植物繁茂而艳丽，给人强烈的视觉感受。然而这么好的风景我们却无暇顾及，因为此行的任务很艰巨，我们要完成商务部援牙买加中国园林的现场勘察和设计，并要使方案获得当地的认可。

Beautiful Caribbean

In December 2008, we flew from the cold Beijing to Jamaica, a beautiful island country in the Caribbean Sea. What came directly in our faces was warm and moist breath with the fragrant of flowers. Lush tropical plants are luxuriant and gorgeous, giving a strong visual feeling to people. However, we had no time to pay attention to the beautiful scenery, because the task of this trip was very difficult. We needed to complete the site survey and design of the Chinese garden in Jamaica, which was sponsored by the ministry of commerce and let the project get the local recognition.

神奇的现场

接下来一个月的时间里，我们沉浸于金斯顿希望公园，现场勘测、现场设计。拂面的微风，大片绵软的草地，两三人才能合抱的大树，还有稍显凌乱的设施，让我们感受到异域风情。更为神奇的是，场地中有一处长满睡莲的百合池塘，池塘里有一座牙买加版图形状的小岛，经常有成群的当地小学生来这里活动，这是他们心中牙买加的象征。保留场地地形和绝大部分现有植被，通过新设计统筹安排剩余空间，植入新的活动内容，这是我们从现场体验得出的结论。

同乐园

在国外建中国园林，既要展示中国园林文化的博大精深，又要把中国文化与当地的生活结合起来。而要想将当地生活融入中国园林之中，对场地的感知非常重要。我们逐渐发现了隐藏在场地中的两个特征并将皇家园林与私家园林的文化特质融入其中，取名“同乐园”。

Magical Scene

We spent the next month immersing ourselves in Kingston Hope Park, surveying and designing the site. The gentle breeze, the soft grass, the trees that only two or three people can embrace, and the slightly messy facilities make us feel exotic. What's even more amazing is that the site features a lily pond filled with water lilies and a Jamaica-shaped island in the lily pond, where groups of local schoolchildren come to play. It is a symbol of Jamaica in their hearts. The terrain of the site and most of the existing vegetation are retained, the remaining space is arranged through the new design, and new activity content is implanted. This is the conclusion we draw from the site experience.

Tongle Garden

To build Chinese gardens in foreign countries, we should not only show the extensive and profound Chinese garden culture, but also combine Chinese culture with local life. In order to integrate local life into Chinese gardens, the perception of the site is very important. We gradually discovered two features hidden in the site and integrated the cultural characteristics of the royal garden and the private garden into it, named "Tongle Garden".

皇家礼仪

场地南侧较开阔，多有大型乔木，散植于大片草地中，围合出几片面积不等的空地，正适于布置活动广场。考虑当地人民举办室外婚礼仪式等的使用需求，自然就联想到了以轴线对称、仪式感很强的广场高台为主要特征的中国皇家园林。设计中以象征天圆地方的方形红色围墙与双层圆形平台为主要景观体，设置南北向与东西向两条轴线，并沿轴线布置几个小型景观节点作为点缀，从而形成了颇具皇家园林特质的方泽院景区。

私家情趣

场地北侧池塘周边，植被繁茂，可用空间尺度大为缩小。于是穿凿亭台楼阁、小桥流水，以小中见大为特色的中国私家园林应运而生。在设计中，我们结合现场植被地形安置了几处小型广场，布置漏窗景墙、棂星门、灵璧石等中国园林元素，使游人步移景异，获得与南部迥异的心理体验，并将百合池塘中牙买加小岛融入其中，私家园林特色的百合池塘景区浑然天成。

Royal Etiquette

The south side of the site is relatively open, with many large trees scattered in large grassland, enclosing several open spaces of different areas, which are suitable for the layout of the activity square. Considering the needs of local people to hold outdoor wedding ceremonies, it is natural to associate with the Chinese royal garden, which is mainly characterized by the square elevation with symmetrical axis and strong sense of ceremony. In the design, the square red wall and the double-layer circular platform, which symbolize the orbicular sky and rectangular earth, are taken as the main landscape body. Two axes are set, north and south, east and west, and several small landscape nodes are arranged along the axes as ornaments, thus forming the scenic spot of Fangze Courtyard with royal garden characteristics.

Private Interest

On the north side of the site, the pond is surrounded by lush vegetation and the available space is greatly reduced. Therefore, the Chinese private gardens with special features expressing the whole through details such as pavilions, small bridges and flowing water emerged at the historic moment. In the design, we placed several small squares in combination with the vegetation and topography of the site, and arranged Chinese garden elements such as leaking window wall, Lingxing gate and Lingbi stone. It enables visitors to enjoy the changing and varied landscape when moving and get a different psychological experience from the south. The Jamaica-shaped small island in the lily pond is integrated into it, and the lily pond scenic spot with private garden characteristics is like nature itself – highest quality.

西昌凉山火把广场

Torch Square in Liangshan Mountain of Xichang

火把广场

西昌所在的凉山州是彝族自治州，彝族部落自称为“火的民族”，崇拜“火图腾”，视“火”为永恒灵魂和希望的象征。彝族传统节日众多，最具标志性的就是“火把节”，其规模之大、内容之丰富、场面之壮观、参与人数之多、民族特色之浓郁，都令人叹为观止。于是，西昌民族文化中心广场以“火把广场”冠名恰如其分。

Torch Square

Liangshan prefecture, where Xichang is located, is the Yi autonomous prefecture. The Yi tribe calls itself "the nation of fire" and worships "fire totem", regarding "fire" as the symbol of eternal soul and hope. There are many traditional festivals of the Yi nationality, the most symbolic of which is the "Torch Festival". The large scale, the abundant content, the spectacular spectacle, the large number of participants and the strong ethnic characteristics of the festival are all breathtaking. Therefore, Xichang national culture center square is appropriately named with "Torch Square".

文化中心

崔愷院士所做的民族文化中心，并没有以建筑为中心，而是以承载当地彝族“火把节”的室外空间为中心，是真正的以“文化”为中心。建筑呈弧形围合出广场，屋顶覆土，消隐体量，与周边的群山环境融合。景观设计也采用了同样的手法，借用建筑施工挖出的土，堆积成叶片状弧形山丘，与建筑呈环抱之势，进一步突显了火把广场的中心地位。

火把广场中心是“永恒之火”雕塑，周边地面上以点状灯源留下星火的痕迹。从主入口进来依次是柱阵广场、中心广场、树阵广场，景观山丘内还有杆影广场星坪花轮、磨秋坪等景观节点。

城市中心

火把广场不仅提供了内涵丰富、层次多样的民族文化展示空间，还营造了民众休闲、雅俗共赏、自由活泼的游憩空间。每到彝历6月24日这天，彝族民众纷纷燃起火把聚集到火把广场，成千上万激情满怀的人们围着熊熊燃烧的篝火载歌载舞尽情欢乐，火把广场成了真正的民族文化中心，真正的城市中心。

Cultural Centre

The national culture center built by Prof. Cui Kai is not centered on architecture, but on the outdoor space hosting the "Torch Festival" of the local Yi nationality, which is truly centered on "culture". The building encloses out the square by arc-shaped, the roof is covered with earth, and the hidden volume is integrated with the surrounding mountain environment. The landscape design also adopts the same technique, using the soil excavated from the construction, piling up into a leaf-shaped curved hill, which encircles the building, further highlighting the central position of the Torch Square.

In the center of Torch Square is the "Eternal Fire" sculpture, and on the surrounding ground there are traces of sparks left by light sources. Coming in from the main entrance is Column Array Square, Central Square, and Tree Array Square; there are also star lawn and floral whorl in Rod Shadow Square, Moqiu Lawn and other landscape nodes in the landscape hill.

City Center

Torch Square not only provides a space for the display of ethnic culture with rich connotation and diverse levels, but also creates a space for people to relax, appreciate both refined and popular tastes, and be free and lively. On June 24 in Yi calendar, Yi people light torches and gather in Torch Square. Thousands of enthusiastic people dance and sing happily around the burning bonfires. Torch Square has become the real center of national culture and the real center of the city.

北川温泉片区

Beichuan Hot Spring Area

羌情
——新北川的
景观起源

2008年的汶川地震牵动了所有人的心，灾后重建也凝聚了各方的力量，我们有幸参与了北川新县城安置房的景观设计，能为灾后重建尽一份力倍感欣慰。由于项目相关事件、人物的特殊性，使得设计并非基于常规的思考模式，而是从重建危机到景观新生意义的探索。反思中，寻求解决问题的策略，回过头来看，所有贯穿设计过程的最重要的线索就是感情，基于北川羌族的羌情成为景观设计的起源。

伤情
——碎屑的刺激，
断裂的北川

北川县隶属四川省绵阳市，羌族人口占少数民族总人口的95.8%，是我国唯一的羌族自治县。在“5 · 12”汶川特大地震中，原北川县所在地曲山镇被列为极重灾区之首，已无原地重建可能。2008年11月初，国务院正式同意北川新县城选址。

现实中坍塌的楼房，断裂的交通，缺腿的桌椅，歪斜的大树，破碎的布娃娃，这些痕迹无一不刺激着我们，令人窒息的伤情扑面而来。

Qiang Affection — the origin of landscape in new Beichuan

The Wenchuan Earthquake in 2008 touched the hearts of all people, and the post-disaster reconstruction also gathered the strength of all parties. We had the honor to participate in the landscape design of the resettlement houses in the new county of Beichuan, and I am glad that I can contribute to the post-disaster reconstruction. Due to the particularity of the events and characters of the project, the design is not based on the conventional mode of thinking, but on the reality from the reconstruction crisis to the new meaning of the landscape. In the process of reflection, strategies are sought to solve the problem. Looking back, the most important clue throughout the design process is emotion. Qiang affection based on Qiang people in Beichuan becomes the origin of landscape design.

Injure Feeling — stimulation of fragments, fractured Beichuan

Beichuan County is part of Mianyang City, Sichuan Province. The Qiang people account for 95.8% of the total population of the national minority. It is the only Qiang autonomous county in China. In the "5·12" Wenchuan Earthquake, Qushan Town, where the original Beichuan County is located, was listed as the most severely affected areas, and there is no possibility of rebuilding in the original site. In early November, the State Council formally approved the location of the new Beichuan County.

In reality, collapsed buildings, broken traffic, desks and chairs without legs, crooked trees, broken dolls··· these marks all stimulated us, and the suffocating injury came directly to our face.

寻情
——重建的彷徨，
未知的北川

北川灾后重建迫在眉睫，要通过援建项目的落实，尽快安置受灾群众，确保人民安居幸福。面对这个艰巨的任务，我们努力寻找重建的答案，但依然彷徨，彷徨于再造的景观空间是否能让北川人民不安定的心获得安全感与归属感。

定情
——文化的回归，
坚定的北川

在现场小会议室的办公桌上，我们将收集到的羌族资料摊开，从羌族人民特有的生活习惯、功能、空间、历史、情感、文化、精神等方面描述设计的布局与思路，我们认为，要从灾后重建的背景下要深挖“羌” 情，紧抓其核心价值特征，要密切关注北川人民的真正需求，这个需求不仅是布局，还有自然、物质、精神、文化的表现与传达，同时不能把景观定义为组织空间的表面艺术，也要关注和保留立于新基址之上场所的原生记忆。

倾情
——空间的激活，
重生的北川

我们从羌族人民的生活入手激活空间，以羌族物质与非物质文化、民风与民俗为主线，对环境中的不同构成元素进行整合和重组，从羌绣、羌族语音音节、碉楼、白石中提取元素，通过艺术化的解构与处理，将其运用到铺装、小品、入口及植物造景当中，强调景观的地域性和民族性。

我们还关注场地的原生记忆，保留现状基址的原生大树及较完整的民居老宅作为景观建筑，同时就地取材，收集拆迁过程中的磨盘、石鼓等农家老物件，并使用当地有特色的毛石、自然石块等作为造景材料，创造一个有意义、有情感的居住环境。

Find Feeling — hesitation of reconstruction, unknown Beichuan

Beichuan post-disaster reconstruction is imminent. Through the implementation of assistance projects, settle the affected people as soon as possible, and ensure that the people live in peace and happiness. Facing this difficult task, we try best to find the answer to rebuild, but still hesitate, wondering whether the landscape space can let Beichuan people's unstable heart to obtain a sense of security and belonging.

Decide Feeling — return of culture, firm Beichuan

On the desk of the small meeting room in the site, we spread out the information we have collected about the Qiang people, and described the layout and thinking of the design from the aspects of the Qiang people's unique living habits, functions, space, history, emotion, culture and spirit. We believe that in the context of post-disaster reconstruction, we should dig deeply into the "Qiang" affection, grasp its core values, and pay close attention to the real needs of the Beichuan people. This need is not only the layout, but also the expression and communication of nature, material, spirit and culture. At the same time, the landscape should not be defined as the surface art of organizing the space, and the original memory of the place should be paid attention to and kept on the new site.

Devote Feeling — activation of space, reborn Beichuan

We start from the life of the Qiang people, activate the space, take the material and intangible culture of the Qiang people, folk customs and folk customs as the main line, and integrate and recombine the different components of the environment. Elements are extracted from Qiang embroidery, Qiang phonetic syllables, watchtower and white stone, and through artistic deconstruction and processing, they are applied to pavement, decorations, entrance and plant landscape, emphasizing the regionality and nationality of the landscape.

We also pay attention to the original memory of the site, and keep the original trees of the current site and relatively complete residential buildings as landscape architecture. At the same time, local materials are collected, such as millstones, stone drums and other old agricultural objects in the process of demolition, and local characteristic rubble stones and natural stones are used as landscape materials to create a meaningful and emotional living environment.

玉树康巴风情商街

Yushu Kangba Custom Commercial Street

海拔 4200 米的“初见”

2010年，我们接到北川之后第二个灾后重建项目，地点在玉树藏族自治州结古镇，是历史上唐蕃古道的重镇。这是个平均海拔在4200米以上，藏语意为“遗址”，被誉为“离天堂最近的美丽地方”。当我们真正走进玉树，初见玉树，才发现本地的自然和人文回归才是设计应有的方向。

连绵的雪山，蜿蜒的河流，广袤的草原，星罗的村庄，成群的牛羊。

自然之初见——感染了。牛羊依托着草原，草原依托着雪山，村庄依托着的河流。

人文之初见——感动了。山口、湖边、寺院、天葬台，无论走到哪里，都是玛尼石堆，特别是废墟中世界上最大的新寨嘉那玛尼石堆尽是虔诚转经修行的玉树人，他们的无常观和生死观使之面对生死处之泰然。

场地之初见——感怀了。地震让结古镇90%的民房倒塌，70%的校舍坍塌，冰冷的扎曲河与巴塘河断断续续流淌在结古镇中央，场地尽显有棱有角、有裂纹的真实。

"First Sight" at 4,200 Meters above Sea Level

In 2010, we received the second post-disaster reconstruction project after Beichuan, located in Jiegu Town, Yushu Tibetan autonomous prefecture, which is an important town on the ancient Tibetan road in history. Jiegu Town is an average elevation of more than 4,200 meters, which means "historic site" in Tibetan and is regarded as the nearest beautiful place to heaven. When we really enter Yushu, and see Yushu at first sight, only to find that the return of local nature and humanity is the direction of the design should be.

Continuous snow mountains, winding rivers, vast grasslands, numerous villages spread over a wide area, herds of cattle and sheep···

The first meet of nature — infected. Cattle and sheep rely on grasslands, grasslands rely on snow mountains, villages rely on rivers.

The first meet of humanity — moved. Mountain passes, lakes, temples, heaven burial platforms, no matter where you go, there are Mani stone pile, and especially in the ruins of the world's largest Xinzhai Jiannamani Stone Pile, there are all the devout Yushu people who are turning scripture and cultivating themselves according to their religious doctrine. Their concept of impermanence and view of life and death make them face life and death calmly.

The first meet of the site — released. The earthquake caused 90% of the residential buildings in Jiegu to collapse, 70% of the school buildings to collapse, and the freezing Zaqu River and Batang River to flow in the center of Jiegu Town on and off. The site is full of edges and corners, with the truth of crack.

穿行 2064 公里的“遇见”

为了设计属于玉树人的居住环境和舒适空间，我们穿越12个县，穿行2064公里，从玉树州到迪庆州，“遇见”康巴藏区的真实。这些真实后续都成为设计语言中自然风貌与材料的显性表达：穿梭在广袤草原上的河流，流淌在河流里的六字真言玛尼石，垭口、湖边的经幡，散落山间草原上的石头房、白藏房、黑藏房……

跨越 1300 多年的“会见”

“初见”文成公主，是在文成公主庙，为纪念文成公主而建，有1300多年历史。峡谷两旁的山上生长着碧绿的松柏、一条蜿蜒的小河清澈见底，虔诚的藏民世世代代悬挂的经幡挂满了文成公主庙边的山头，经幡摇动，像一条长虹，将汉、藏人民紧紧连在一起。

“会见”文成公主，是希望将唐蕃古道从内地一直往西到雪域高原的汉藏团结故事线及玉树人对文成公主的情感线融入设计中，同时也见证玉树的浴火重生，于是文成公主进藏的故事转换为设计的剧本——呈现。

"Acquaintance and Encounter" by Travelling 2,064 km

In order to design the living environment and comfortable space belonging to Yushu people, we crossed 12 counties and traveled 2,064 kilometers from Yushu Prefecture to Diqing Prefecture to "encounter" the reality of Kangba Tibetan area. All of these realities have become the explicit expression of natural features and materials in the design language: the river that shuttled across the vast grassland, the Mani stone with six-character mantra flowing in the river, the prayer flags in mountain passes and lakes, the stone house, white house and black house scattered on the mountain grassland···

"Meeting" Spanning More Than 1,300 Years

First meeting with the Princess Wencheng is in the Princess Wencheng Temple, where was built to commemorate the Princess Wencheng with more than 1300 years of history. The mountains on both sides of the canyon are green with pine and cypress, and a winding river is clear to the bottom. Prayer flags hung by devout Tibetans from generation to generation are all over the hill beside Princess Wencheng's Temple. The prayer flags like a long rainbow, holding the Han and Tibetan people together.

The meeting with Princess Wencheng, is hoping to integrate the story line expressing Han and Tibetan unity in the ancient Tibetan road from the mainland of China to the west snowy plateau and the emotional line of Yushu people to Princess Wencheng into the design, and also to witness the rebirth of Yushu, so the story of Princess Wencheng's journey into Tibet transforms into a design script and presents.

回归2012年的“觐见”

从“初见”到“遇见”再到“会见”，都在告知有一种美丽，叫玉树。因此设计立足于自然风貌的抽象化及景观的叙事化。将自然、人文两条线索纳入景观方案的构思当中。以自然为线索，基于大地的设计语汇和风格，将延绵的山峰、流淌的河流、堆砌的玛尼石进行抽象处理，以景观铺地、置石、台阶等形式塑造场地自然基底，使得场地内的建筑成为基底中生长出的有机元素。其次以人文为线索，将文成公主进藏的故事融入各个景观节点，通过空间对景、景观雕塑、象征图案等要素将故事情节步步推进，将聚缘、赏纳、竞曲、融合、祝福、升华、幸福等场景通过有张有弛的故事情节步步展开。最终设计回归场地本身，对周边场地进行视线关系分析，强化了场地看与被看的关系，保持视线与空间的连续性，同时将视线分析作为预留视线通廊的主要依据，有效地把周边场地中有利的景观纳入场地整体景观范畴，实现空间物化与精神意象的升华。

"Formal Visit" Back to 2012

From the first sight to the acquaintance and encounter, and to the meeting, are told there is a beauty, called Yushu. Therefore, the design is based on the abstraction of natural scenery and the narration of landscape. Natural and cultural clues are incorporated into the design of the landscape plan. Taking nature as the clue and based on the design vocabulary and style of the earth, the abstract treatment of the lingering mountains, flowing rivers and piled Mani stones is carried out. The natural basement of the site is shaped in the form of landscape paving, stone setting and steps, so that the buildings in the site become organic elements growing out of the basement. Secondly, taking humanity as the clue, the story of Princess Wencheng's entry into Tibet is integrated into each landscape node. Through the opposite landscape in the site, landscape sculpture, symbolic design and other elements to advance the story step by step, through the story plot with tension to expand gathering luck, appreciating reward, competing songs, integrating, blessing, sublimating, happiness and other scenes step by step. Finally, the design returns to the site itself, analyzes the line of sight relationship of the surrounding sites, strengthens the relationship between the site seeing and being seen, and maintains the continuity of line of sight and space. At the same time, the line of sight analysis is taken as the main basis to reserve the line of sight corridor, and the favorable landscape in the surrounding site is effectively incorporated into the overall landscape category of the site, so as to realize the sublimation of spatial materialization and spiritual image.

3. 古韵新做

Ancient Rhyme with New Advantages

万科紫台

Vanke Zitai

传统与现代的思考

随着国内景观设计的蓬勃发展，越来越多的人开始思考如何构建景观理论体系并探索设计方法。思考中不能回避的是对传统和现代的理解，对中国古典造园理论和西方现代景观体系的对比和取舍。虽然业界争论很多，但不排除走融会贯通的路线。万科紫台项目给我们提供了这样一个思考和实践的机会，让我们初步总结出古韵新做的景观策略，算是对传统与现代结合的设计方法的一家之言吧。

万科紫台项目位于北京城西部，东边与西四环很近，北侧离长安街不远。地理位置决定了两种主要客户群：一是具有大院情结的人，二是追求西贵气质的人。西部地块所居住的人群大部分生活在军区大院、机关大院和学校大院。他们在日益紧张的都市生活中，越来越渴望儿时大院生活的宁静与悠闲；而居住在长安街边、皇城脚下的他们对于西贵气质的追求，对于皇家文化的探究又是无止境的。将皇家的气质与文化、大院生活的氛围带入新的景观设计之中，让人们既能在现实生活中体味怀旧情怀，又能感受到现代时尚氛围，就成为紫台项目景观设计的最初理念。于是我们提出了古韵新做，我们认为，古韵新做首先要理解古韵，然后再新做。古韵新做包含三个层次：取其意，得其法，用其形。

Traditional and Modern Thinking

With the rapid development of landscape design in China, more and more people begin to think about how to construct the landscape theory system and explore the design method. What can't be avoided in thinking is the understanding of tradition and modernity, the comparison and choice between Chinese classical garden theory and western modern landscape system. Although there are a lot of debates in the field of landscape, do not rule out the line of an achieve mastery through a comprehensive study. Vanke Zitai provides us with such an opportunity to think and practice, lets us preliminarily summarize the new landscape strategy of the ancient charm; can be regarded as idiosyncratic views of the design methods after the combination of traditional and modern.

Vanke Zitai is located in the west of Beijing, near the west fourth ring road in the east and Chang 'an Avenue in the north. Geographical location determines two main customer groups: one is the people with the emotion of the courtyard; the other is the people who pursue of western noble temperament. Most of the people living in the western plot live in the military compound, government compound and school compound. In the increasingly tense city life, they are more and more eager for the quiet and leisurely life in their childhood. And the people living in the streets of Chang 'an, at the foot of the imperial city, pursue of western noble temperament, and the endless exploration of royal culture. To bring the royal temperament and culture, the life atmosphere in the courtyard into the new landscape design, so that they can not only feel the nostalgia in real life, but also feel the modern fashion atmosphere, became the initial concept of the landscape design of the Vanke Zitai. So we put forward the idea—ancient rhyme with new advantages, we think, it needs to understand the ancient rhyme at first, then adds the new advantage. Ancient rhyme with new advantages contains three levels: take its meaning, get its method, and use its form.

意趣捕捉——取其意

古韵新做首先要合理地选择古韵，充分理解古韵，这是一个痛苦的过程。经过多个思路比较，我们最终选择了引入国学的理念。将传统国学的七门教学科目（礼、乐、律、射、御、书、数）引申发展，用现代景观语言去加以阐释，以此形成了万科紫台独特的景观气质——古雅中彰显时尚，现代中古韵犹存。这就是古韵新做所追求的意境，从这个意义上说，万科紫台也将传统造园追求意境的精髓进一步传承了下来。

接下来的工作是如何将国学与庭院空间结合起来，满足居住者的大院情结。根据万科紫台建筑布局围合成三个院落的不同属性，将国学的七艺进一步分解，以适应不同使用功能的要求，创造各具特色的庭院意趣。中间的大院作为示范区，公共性、展示性较强，我们用礼、乐、律的主题去表现皇权的礼仪性，将其定义为龙院。南北两个院落以居住为主，我们将其定义为一文一武两个院落，文院用书、数去表达，武院用射、御去表达。这样一来，国学的七艺巧妙地与院落空间结合起来，每个院落都有了不同的意境，为古韵新做的进一步展开找到了线索。

Capture of Interest and Charm — take its meaning

Ancient rhyme with new advantage needs to choose ancient rhyme reasonably above all, understands ancient rhyme adequately, and this is a painful process. After comparisons with several thinking, we finally chose to introduce the concept of sinology. The seven teaching subjects of traditional sinology (ritual, music, law, archery, imperial art, calligraphy and mathematics) are extended and developed, and interpreted with modern landscape language, thus forming the unique landscape temperament of Vanke Zitai — showing the fashion in the ancient elegance and the ancient rhyme in the modern. This is the artistic conception pursued by the ancient rhyme with new advantage. In this sense, Vanke Zitai has further inherited the essence of the artistic conception pursued by the traditional garden.

The next work is how to combine sinology with courtyard space to meet the courtyard emotion of the habitant. The different properties of the three courtyards are enclosed according to the architectural composition of Vanke Zitai, and the seven arts of sinology are further decomposed, to meet the requirements of different using functions and create the distinctive courtyard interest. As a demonstration area, the courtyard in the middle has a strong public and display nature. We use the theme of ceremony, music and law to express the ceremonial nature of the imperial power and define it as the dragon courtyard. The two courtyards in the north and south are mainly residential. We define them as two courtyards in the form of literature and martial arts. The literature uses calligraphy and mathematics to express, while the martial arts uses archery, imperial art to express. In this way, the seven arts of sinology are skillfully combined with the courtyard space; each courtyard has a different mood, and finds a clue for the further development of the ancient rhyme with new advantage.

手法选择——得其法

每个院落的主题确定后，用什么手法将国学七艺与景观元素取得关联，成为思考的重点。古韵新做不是照搬古典园林，而是以现代的格局、现代的手法去诠释古韵。在示范区的龙院，我们用下沉的方形小广场去表现礼的规矩，用圆形的高台去展现乐的柔和，并附以叠水和回音壁的红墙，进一步表达乐的魅力。在小广场和高台之间用象征律的御道加以连接，拾级而上，更显皇家礼仪。

南北文院和武院则以更具生活情趣的手法去表达主题。文院中的砚池体现了文人的生活场景，曲水流觞表达了古典园林的清雅，再附以诗词歌赋的雕刻，更表达了文人墨客的生活追求。武院用古代的车辙痕迹以及车马的雕塑去表现金戈铁马的射御之术，从而唤起人们对武学的崇尚。

Choice of Technique — get its method

After the theme of each courtyard is determined, how to associate the seven arts of sinology with landscape elements becomes the focus of thinking. The ancient rhyme with new advantages is not to copy the classical garden, but to interpret the ancient rhyme with the modern pattern and modern technique. In the dragon courtyard of the demonstration area, we use a sunken tetragonal square to show the rules of etiquette, and a round high platform to show the softness of music, and with the folding water and the red wall of echo wall, to further express the charm of music. The small square and the high platform are connected by a road which is a symbol of the law; step up, more royal etiquette is showed.

The north and south courtyard of literature and the martial arts use the methods with more life to express the theme. The ink-stone pond in the literature courtyard reflects the life scene of the literati, the winding streams express the elegance of the classical garden, and the sculpture of poems and songs expresses the pursuit of life of the literati. The martial arts courtyard used ancient ruts and statues of chariots and horses to express the art of imperial shooting and riding, thus arousing people's admiration for martial arts.

细节营造——用其形

古韵新做除了意境营造和手法表达外，还应注重细节的研究，所谓细节决定成败。万科紫台的景观设计，力图将传统造园的理法体现在每一个细节上，但细节的材料、形态又是现代的。景观营造充分考虑对景、框景、借景的关系，在视线的开闭转合等方面精心研究。园路曲径通幽，步移景异；七艺主景周边还有小的节点，园中有园，景中有景；水从圆台的龙源中流出，顺阶而下，形成一潭扇形鱼池；会所前的方形水池中条石结合喷泉，现代时尚；植物主次搭配，随景而生；置石顺应古法，自然巧妙；廊架、景墙等在简洁现代的形态中配有古典图案的漏窗、龙纹等装饰，凸显情趣和品位。

Details Construction — use its form

In addition to artistic conception construction and expression, ancient rhyme with new advantages should also pay attention to the study in details; the so-called details determine success or failure. The landscape design of Vanke Zitai tries to embody the traditional garden theory in every detail, but the material and form of the details are modern. Landscape construction fully considers the relationship between the opposite scenery, enframed scenery, and borrowed scenery, and carefully studies in the aspect of the line of sight opening, closing, and turning. The winding path of garden road leads you to a secluded place, with different views; there are small nodes around the main landscape of seven arts, garden in garden, scene in scene; the water from the dragon source of the round table flows down the steps to form a fan-shaped fishpond; the square pool in front of the club combines the stone with the fountain, which is modern and fashionable; plants are in primary and secondary collocation, existing with the scene; stones are comply with the ancient law, natural and ingenious; gallery frame, landscape wall and so on are in concise and contemporary configuration, and are equipped with the adornment such as the leakage window of classical design, dragon grain, protruding shows appeal and grade.

南长安街壹号

No. 1, South Chang'an Avenue

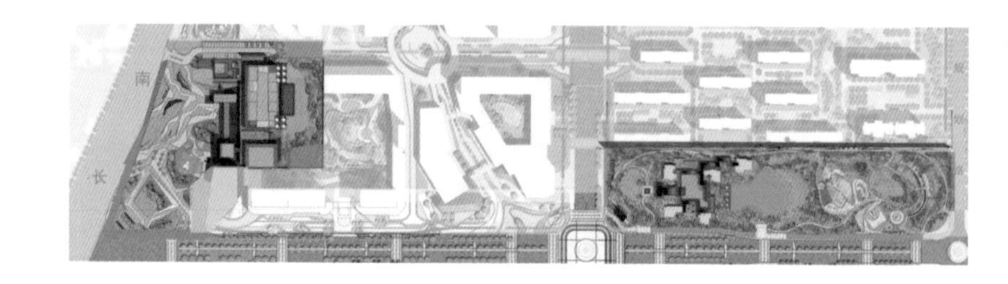

风雅与惊艳

由于此项目是业主进军古都西安的第一个项目，因此拿到这个项目初始，我们的团队是既兴奋又有些踌躇，现代楼盘与古都西安，使得这个项目的笔墨拿捏非常微妙。

之于悠远：项目选址旧称韦曲，细查其背后的历史，正是“千年城南书香，一席钟灵毓秀”之地，多少文豪墨客曾建馆于此，寄情山水，谈笑鸿儒，留下许多千古名篇待后人品鉴，给场地烙下独特的风雅印记，切中场所文化气质是这次命题作文的题眼之一。

之于现实：项目作为地产集团进军西安的重要高端项目，需要打出一记漂亮的商业惊叹号，这也使得项目气质要迎合世俗需要，满足其作为最流行商品的惊艳特质。

如果是纯粹文化类景观或者纯粹商业类景观，设计的发力点都非常明确，而本案雅俗都要兼顾又都不可用力过猛，如何权衡笔墨成了设计重点。

Elegant and Stunning

Since this project is the first project for the owner to enter the ancient capital Xi 'an, our team is both excited and hesitant when we got this project in the initial stage. Modern real estate and ancient capital Xi 'an make this project's identity very delicate.

To the distant: the project site was formerly known as Wei Qu. Carefully checking the history behind, it is a place that "book fragrance in the south of the city for a thousand years, endowed with the fine spirits of the universe". How many great writers and poets have built their houses for activities here, abandoned themselves to nature, laughed and talked freely with knowledgeable people and left many famous works to be tasted by future generations, branded a unique elegant mark to the site. The cultural temperament is one of the eyes of this thesis composition.

To the reality: as an important high-end project of real estate group entering Xi 'an, the project needs to make a beautiful commercial exclamation point, which also makes the project temperament to meet the secular needs and satisfy its amazing characteristics as the most popular commodity.

If it is a pure cultural landscape or a pure commercial landscape, the force point of the design is very clear. But for this project, both elegance and vulgarity should be taken into account and should not be too strong, how to balance the ink has become the focus of the design.

不负长安不负卿

我们执着同时对场地文脉和项目展示负责的态度，翻阅了大量史籍资料，发现旧时此地（韦曲樊川一带）因其独特的地理位置，背靠终南山绝景，少了一份皇城腹地的禁锢与约束，终以世外高人之姿，悠游于出世入世之间，正是如此清雅洒脱的气质，吸引文人士大夫悉数前往，建馆立业。这种郊野文人别馆，既有皇城院落的尊贵，又有世外桃源的隐谧，是一种“兼顾大融”的气质，它应当是尊贵典雅但不刻意张扬，简洁质朴但不过分简陋的。

场地这种介乎出世与入世的中隐之姿，正是我们努力找寻的破题之法。我们把这样的气质用设计语言表达出来，于是便有了这样一个设计：你会看到中正的院落空间逐渐向郊野公园过渡；会看到考究的尺度精致的搭接，又会看到貌似漫不经心的原生堆砌；会看到精心修剪的桂花、银杏，又会看到肆意蔓延的菖蒲、芒草；会看到精致的铜活，又会看到粗犷的竹木；会看到两种气质的转换融合。这种“半露书卷，半隐繁华”的设计，是我们同时对古都西安和现代楼盘的答复，我们要努力做到“不负长安不负卿”。

这样一个空间和气质的过渡，需要一条线索串联，所以我们用设计语言营造了一个“找寻”的故事，展现这样一组场景：绕开层峦清障，步入玄开山门，踏进湖池如镜的优雅精舍，于芳林草地间信步游走，最终寻到了质朴返真的山水园林。

Live up to Chang'an and Live up to You

Determined to be responsible for both the context of the site and the presentation of the project, we consulted a large number of historical documents, and found that this place (in the area of Weiqu and Fanchuan) in the old days was less confined and restrained by the imperial city hinterland because of its unique geographical location and its back to Zhongnan Mountain. Finally, with the appearance of the world's outstanding people, leisurely travel between coming in and going out the world . It is so elegant and free temperament, attracting all scholars and literati to set up a home and establish a business. This kind of countryside literati house, has both the dignity of royal court, and the seclusion of the arcadia, is a kind of "give consideration to overall situation" temperament. It should be noble and elegant but not deliberately make widely known, simple but not too crude.

The hidden appearance of the site between coming in and going out the world is the method we are trying to find. We express such temperament in design language, so there is a design: you will see a gradual transition of structured courtyard space towards country parks; will see the exquisite and finely scaled lap joint, and will see the seemingly careless original stack; will see the well-pruned osmanthus ginkgo, and will see the spread of calamus miscanthus; will see the delicate copper work, will see the rough bamboo; will see a shift and combination in temperament. This kind of "half open scroll, half hidden prosperity" design, is our answer to the ancient capital of Xi 'an and modern real estate at the same time, we should strive to do "live up to chang 'an and live up to You".

Such a transition between space and temperament need a thread to connect, so we use the design language to create a "search" story, to show a set of scenes: bypass layer upon layer of mountains, step in the half-open mountain gate, set foot in elegant house with a lake like a mirror, take a leisure walk among the trees and grassland, and finally find the place return to the real landscape garden.

设计的中庸之道

首先，在这里要还原“中庸”之意，词语出自《论语 · 庸也》：“中庸之为德也，其至矣乎”，意为不偏不倚。中庸一词从来都带有褒义色彩，不是后世理解的折中妥协，而是一种天人合一的哲学观，反映了天性与人性的合一，理性与情感的合一。设计师很难碰到一个项目，能完成得出尘脱俗、仙风道骨，中间势必杂糅了用户的需求和其他诸多限制条件，设计师需要做到满足自我的同时让多方满意。在这个项目中，我们逐渐摸索到了设计的中庸之道，同时实现了客户认可和自我满足。

Moderation in Design

First of all, here to restore the meaning of "moderation", it is from the Yongye in Analects of Confucius: "moderation is virtue, it is the most supreme", means impartial. The word "moderation" has always had a positive connotation. It is not a compromise understood by later generations, but a philosophical view of the unity of nature and man, reflecting the unity of nature and humanity, reason and emotion. It is hard for a designer to meet a project, which can be completed free from worldly things, the process is bound to be mixed with the user's needs and many other restrictions, the designer needs to be satisfied with their own at the same time. In this project, we gradually explored the moderation of design, while achieving customer recognition and self-satisfaction.

十堰四方新城

Shiyan Sifang New City

“不矫情，悠游自在”

“那年从厂子出来了，和大伙一块干起来，还是想做点不一样的事儿”“确实要挣钱，也挣不完呀”“不能太矫情，悠游自在，做点自己喜欢的”。这是十堰四方新城开发者跟我们说的话。弃工从商的经历他历历在目，甘苦之后之所以选择四方山周边作为他的项目之地还是希望还自己一个隐士夙愿，所谓“小隐隐于林，大隐隐于市”。倏地，眼前的“随园老人”便活生生地跃然而出，于是我们便一拍即合，取袁枚“随园诗话”中“随园”一词，作为他项目的雅名，“随园”随缘而生，希望借四方山这片山林还他一个心灵的归宿。

"Don't be melodramatic, be at ease"

"I came out of the factory that year, worked together with friends, and wanted to do something different" "You really want to make money, but you can't make it all" "Don't be too melodramatic, be at ease, do something you like". This is what the developer of Shiyan Sifang New City told us. He has vividly remembered the experience of abandoning work and going into business. After the hardships, he chose the surrounding area of Sifang Mountain as the site of his project and he hoped to return his long-cherished wish of a hermit. The so-called "A little dull in the forest, a great dull in the city". Suddenly, the "old man who owned the garden" in front of our eyes jumped out alive, so we hit it off and took the word "Suiyuan" in Yuan Mei's "Suiyuan Poetry Talks" as the name of his project. "Suiyuan" is built with fate, hoped to borrow this mountain forest of Sifang Mountain to return him a spiritual home.

何为古韵 新作何为

“古韵”还是要品出来的才好，太直白了也就没了味道。既是定了随园的诗话作为追求，便要研习其中之“古”方得“随园之韵”。清代性灵诗派的倡导者袁牧有诗云：“造屋不嫌小，开池不嫌多；屋小不遮山，池多不妨荷。游鱼长一尺，白日跳清波；知我爱荷花，未敢张网罗”。诗人在其随园的门楹上题联：“放鹤去寻山鸟客，任人来看四时花”。四方新城“随园”落于山地之中，仅有西侧紧邻市政道路，既有现代交通的畅达，又不少优越的自然生态环境，依山建筑风格是抽象了的白墙黛瓦，现代古风山野之韵由此而生。诗画古韵的山地居所，融于山形的起伏，层叠的树木，潺潺的溪流，正是我们要营造的氛围。

What is Ancient Rhyme, How to do the New Work

"Ancient rhyme" had better to be tasted out by people, and too straightforward do not have taste. Since we decided to follow the garden's poetry as the pursuit, we should study the "ancient" in it to "follow the garden's rhyme". Yuan mu, an advocate of the Xingling School of poetry in the Qing Dynasty, had poetry: "Don't mind the small house, but the more ponds there are, the happier I am. The small house will not block out the mountains and the ponds can be used to plant more lotus. Swimming fish grow one more foot long, and they jump in the pond during the daylight. People know I love lotus so they don't dare to cast the net to catch." The poet wrote a couplet on his garden gate: "Let the crane go in search of the visitors, and let the visitors come and see the four-season flower." "Suiyuan" in Sifang New City falls in the mountains, with only the west side adjacent to the municipal road, which not only has the smooth modern traffic, but also has many superior natural ecological environments. The architecture style based on the mountains is abstract white walls and black tiles, which gives rise to the charm of modern ancient style and wild scenery. The mountain dwelling with ancient rhyme of poetry and painting, blending in the undulations of mountain forms, with the overlapping trees and the babbling streams, is the atmosphere we want to create.

新作之困，自然解围

山地建造住宅群落不可避免地要与原生山体产生交融和碰撞，园林景观的营造也受山地条件限制出现迂回路转、高低纷杂和山石修葺的矛盾。

迂回路转之困，我们用婉转拾趣解围。漫步于自然山路的婉转，院落和小场地借山形迂回之变跳跃着，一条依仗山墙一侧的主轴贯穿了每个跳跃空间，主轴收放交替，随时随势地创造了丰富的居民活动空间。

高低纷杂之困，我们用望岳闻声解围。山地院子视线层次多样，让人体验到不同的景致，远可见山岳，中可俯瞰层叠院落，近可游院中美景。自古山水不分，悠游山趣寻水声。主轴中心空地作四水归堂，水从中心水池向各台地院落叠落而去，水声于每一个院落产生“虽有高台异远近，唯有活水天上来”的意境。

山石修葺之困，我们效法自然解围。因为建筑施工原因，原本墨绿浓荫的山体局部露出岩石，是覆绿还是保持现状，我们走进场地发现绿色基底上隐约可见的山石体现出另一种原生态的美。于是，我们在适当地恢复植被外，大多保留了现状的岩石山体，并取材山石修葺了层层叠叠的挡土墙，工艺效法自然，利用石材的无勾缝拼贴，找到了人工砌筑和自然山石的平衡点。

随园景观的营造在一定程度上满足了内心情感的外化，探讨了人与自然的融合之法，正是出于自然、归于自然的道家心性使得融于山林的现代居住环境带有一种古典诗画般的中国神韵。

Dilemma of New Work, Solve by Nature

Residential communities built in mountainous areas inevitably blend with and collide with the original mountains, and the construction of garden landscape is also restricted by mountain conditions, resulting in roundabout turns, conflicts between high and low and mountain stone repair.

The trap of roundabout turns, we use mellow and full to solve. Rambling over the mellow and full of natural mountain roads, courtyards and small sites use mountain forms to skip around, and a main axis along one side of the gable runs through each jumping space. The main axis is not melodramatic; it is alternating, at any time and situation with the potential to create a rich living space.

The conflicts between high and low, we use seeing the mountain and hearing the sound to solve. The mountain courtyard has a variety of views, allowing people to experience different scenery. Mountains can be seen in the distance, the stack-up courtyard can be overlooked nearly, and the beautiful scenery of the courtyard can be visited closely. Since ancient times, the landscape between mountains and rivers is not divided; people can find the interests of mountains and search water sound at the same time. The central space of the principal axis is used for four-water returning to the hall. The water flows from the central pool to the courtyard of each platform. The sound of water generated the mood in each courtyard— "although there is a different distance, only living water from the sky".

The trap of mountain stone repair, we follow the example of the nature to solve. Because of the construction, the original dark green and shady mountain is partly exposed to rock, whether covered with green or kept as it is, we walked into the site and found that the faint mountain rocks on the green basement reflect another kind of original ecological beauty. So, we kept most of the existing rock mountains, except for the proper restoration of vegetation. And we repaired the layered retaining walls from mountain rock. The technology imitates the nature, and finds the balance point between artificial masonry and natural mountain stone by using the stone collage without pointing.

With the construction of the "Suiyuan" landscape, the externalization of the inner feelings is satisfied to a certain extent, and the fusion of man and nature is discussed. It is the Taoist spirit that comes from and back to nature gives the modern living environment in the mountains a Chinese charm like classical poetry and painting.

4. 喻礼于境
"Rites" and "Environment"

雁栖岛APEC会议中心
APEC Conference Center of Yanqi Island

传统的“礼”与当代的“境”

中国是传统的礼仪之邦，我们很多项目都需要以今天的视角及当下的情怀去解读中华文化的礼制精神和艺术魅力。雁栖湖国际会都APEC核心岛入口及南广场的设计就是一个典型的案例，我们用礼仪形制营造空间格局及行为动线，用情与意延展中华文化的艺术魅力，探索“礼”与“境”的融合，让文化回归当下，让景观溯源意境，让设计重新找到文化的认同。

Traditional "Rites" and Contemporary "Environment"

China is a country of traditional rites, and many of our projects need to interpret the spirit of rites and artistic charm of Chinese culture from today's perspective and current feelings. The design of the entrance and the south plaza of the core island of the APEC Conference Center of Yanqi Island is a typical case. We use the form of etiquette to create the spatial pattern and the dynamic line of behavior, extend the artistic charm of Chinese culture with emotion and artistic conception, explore the fusion of "rites" and "environment", let the culture return to the present, let the landscape trace the artistic conception, let the design find the cultural identity again.

空间文化主义与文化田园思想

“礼”是传承中国多年的教仪，中国人也以其彬彬有礼的风貌而著称于世。礼仪文化根植于中国的个人、社会、国家各个层面,对历史产生了深远的影响，礼仪文化也不断向各领域发展、延伸。其在空间营造上的特质主要表现在空间格局、空间体验和文化元素上。我们通过空间开合、轴线营造、借景障景等方式使山水关系、场地与建筑关系等形成礼仪序列、加强轴线的仪式感以及多种空间交叠的体验，使得礼仪功能的要求得以满足，空间的序列也得以展开。

“境”是中国园林追求的目标，“景”是具体的、直观的，而“境”是对“景”的更完善的表达，是超越“景”的更高境界。天人合一作为中国人归纳出的人与自然的至上关系,表达出深入人心的文化田园思想，我们在由“景”到“境”的实现过程中，把思想、心境通过可视的景观元素加以传达,让这一切景物变得可以感受，可以玩味，可以思索。借汉阙、御冕、花格为形象载体，通过抽象、提炼、借喻、解构、重组等设计手法，凝练出植根于中国传统文化内在结构的景观语言，传承与延续中国传统文化的精髓。

Spatial Culturalism and Cultural Pastoral Thoughts

"Rites" is a tradition that has been passed down in China for many years. Chinese people are also famous for their politeness. The etiquette culture is rooted in the individual, social and national levels of China and has exerted a profound influence on history. The etiquette culture also develops and extends to various fields. Its characteristics in space construction are mainly manifested in spatial pattern, spatial experience and cultural elements. We express the etiquette sequence experience through relationship between mountains and rivers, the relationship between site and architecture by opening and closing space, constructing axis, borrowing scenery and blocking scenery. The sense of ceremony of the axis and the experience of multiple overlapping spaces enable the requirements of etiquette functions to be met and the sequence of spaces to be expanded.

"Environment" is the goal pursued by Chinese gardens, "landscape" is specific and intuitive, while "environment" is a more perfect expression of "landscape" and a higher realm beyond "landscape". As the supreme relationship between man and nature summed up by the Chinese, the unity of man and nature expresses deeply rooted cultural pastoral thoughts. In the realization process from "landscape" to "environment", we convey thoughts and moods through visual landscape elements, so that all these scenes can be felt, can be played, can be thought. With the image carrier of Han Que, royal crown and flower lattice, the landscape language rooted in the internal structure of Chinese traditional culture is condensed through abstraction, refinement, metonymy, deconstruction and recombination, inheriting and continuing the essence of Chinese traditional culture.

奥林匹克下沉花园

Olympic Sunken Garden

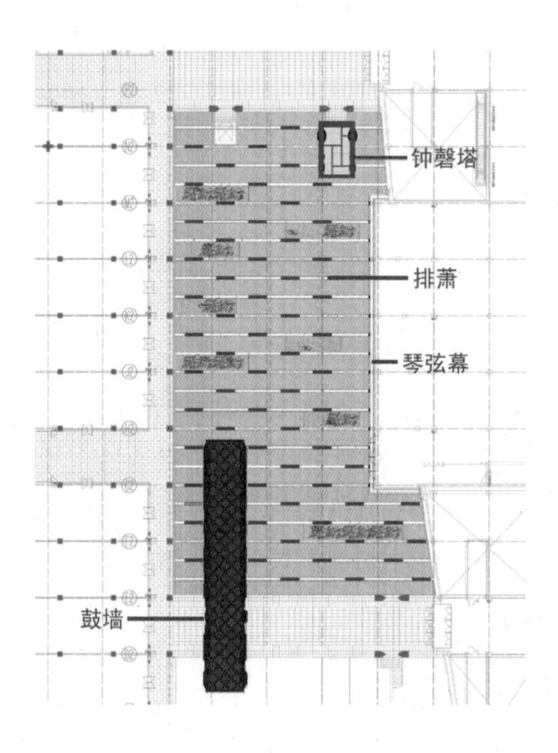

礼乐重门

“2008 北京奥运”是古老的中华文明和现代奥林匹克精神的首次融合，秉承着绿色奥运、科技奥运、人文奥运三大理念，北京这座以紫禁城和四合院闻名于世的历史文化名城注定要再次见证新的奇迹。

奥林匹克中心区在北京中轴线的北延长线上，中轴线东侧的7个下沉庭院成为奥运庆典期间集中展示中华文化精粹的场所。我们负责的其中一个院子东临龙形水系下的地下商业用房，西靠地铁奥运支线出入口，南北联通前后院子，是一个交通和商业交汇的空间。

崔愷院士将院落主题定义为“礼乐重门”，试图表达中国传统的“礼乐文化”，用以欢迎来自四面八方的游客。我们在这个定位下展开思考，尝试用景观语言营造庆典和礼宾的环境氛围。

Heavy Door of Rites and Music

The 2008 Beijing Olympics Games is the first fusion of the ancient Chinese civilization and the modern Olympic spirit. Adhering to the three concepts of green Olympics, high-tech Olympics and humanistic Olympics, Beijing, a historical and cultural city famous for its Forbidden City and quadrangle courtyards, is destined to witness a new miracle again.

The Olympic central area is on the north extension of the central axis of Beijing, and the seven sunken courtyards in the east of the central axis serve as a focal point for showcasing the essence of Chinese culture during the Olympic celebration. One of the yards we are responsible for is adjacent to the underground commercial buildings under the dragon-shaped water system in the east and the entrance and exit of the subway Olympic branch in the west. And its north and south connect front and rear courtyard, is a traffic and commercial intersection of the space.

Academician Cui Kai defined the theme of the courtyard as "heavy door of rites and music", an attempt to express the traditional Chinese "culture of rites and music" and to welcome tourists from all over the world. We think in this position, trying to use landscape language to create an environment of celebration and hospitality.

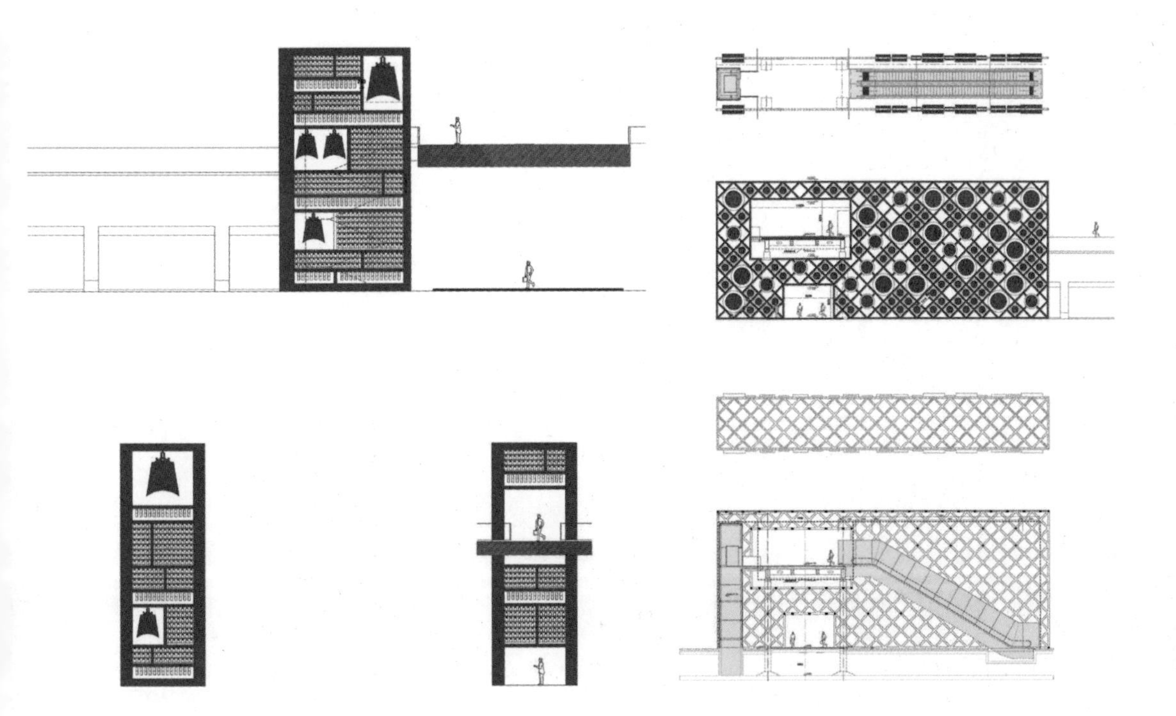

礼乐新用

以传统迎宾待客古乐礼器中的“钟”“磬”“鼓”“萧”“弦”等为原型，充分利用下沉花园中的室外扶梯、电梯、楼梯等交通空间，我们创作了一系列新用的礼乐之器，包括自动扶梯外以大小不同的鼓组成的“鼓墙”，电梯间外悬挂钟磬的“钟磬塔”，庭院内作为灯具使用的一排“排箫灯”，以及商业建筑玻璃窗外的“琴弦幕”。这些装置既有实用功能又有艺术效果，配以典型的国红、金黄、铜青等中国色彩，掩映于竹林之中，充分展现儒雅风骨与大国威仪相并重的礼乐文化。

New Use of Rites and Music

Taking the traditional "bell" "chime" "drum" "flute" and "string" in the ancient music ritual of welcoming guests as the prototype, the outdoor escalator, elevator and staircase in the sunken garden are fully utilized. We created a series of new ritual instruments, including the "drum wall" outside the escalator composing with drums of different sizes,the "bell chime tower" that hangs the bell chime outside the elevator, a row of "pipe lamps" used as lamps in the courtyard, and a curtain of strings outside the glass windows of commercial buildings. These installations have both practical functions and artistic effects, with typical Chinese colors such as red, gold and bronze, and are set in the bamboo forest, fully displaying the ritual and music culture that pays equal attention to the elegance and dignity of a big country.

科技互动

为了体现大众参与、体验、互动的奥运精神，钟、磬、鼓、琴、箫等景观装置小品均采用工业设计技术。参观者在行进中可以通过击打或拨动等方式进行演奏，模拟声效技术，使参观者体验到参与其中的行为乐趣。我们还装置了可以在风的作用下产生连动声效的“风铃”“排箫”等空间构成元素，让自然发出声音。正是通过多方面的“互动式场景”设计，使得整个下沉广场空间得以激活，体现出与传统文化相交融的科技奥运精神。

Interaction of Science and Technology

In order to embody the Olympic spirit of mass participation, experience and interaction, the landscape devices such as bell, chime, drum, musical instrument and flute adopt industrial design technology. During the procession, visitors can play by hitting or plucking, simulating sound effect technology, so that visitors can experience the fun of participating in the performance. We also installed the "wind chimes" "panpipe" and other spatial components that can produce continuous sound effect under the action of the wind, so that the natural sound can be generated.It is through the design of various "interactive scenes" that the whole sunken square space can be activated, reflecting the high-tech Olympic spirit blended with traditional culture.

德胜尚城
DE Sheng Shang City

过往与更新

德胜门的西北角，曾经是胡同小院密布的地区，虽然不如城里四合院那么讲究，但承载的市井生活却大同小异。因为靠近什刹海，这里曾是北京三大冰窖之一，直到1960年代还延续着凿冰窖储的习惯。现如今仅剩下冰窖口胡同的名字聊以慰藉。我们要设计的场地东西宽约100米，南北长约210米，拆迁之前，这里有6条胡同，生活着152户人家，那种“天棚鱼缸石榴树，先生肥狗胖丫头”的生活场景如今要被一组现代化的多层办公建筑所取代。

尊重与寻找

进入21世纪，北京城更新的速度明显加快。有感于老城风貌的逐渐褪色，设计师总希望能留下点回忆，保留些对以往生活的尊重。为了尊重德胜门城楼，崔总（崔愷）在设计建筑时，在地块内做了一条斜街，正对着德胜门，斜街两侧的建筑不规则地围合出几组院落，形成了新的胡同和院落肌理。这种做法为景观设计奠定了基调，更多人情味的构建要靠景观设计去实现。如何延续这种理念，又如何保留城市记忆呢？我们开始了寻找。

Past and Update

The northwest corner of Desheng Gate used to be a densely populated area of hutongs. Although it is not as elegant as the quadrangle courtyard in the city, the community life of the city is similar. Because of its proximity to Shichahai, it used to be one of the three major ice cellars in Beijing, and the custom of digging ice cellars to stockpile continued until the 1960s. Now only the name of Bingjiaokou Hutong is a consolation. The site we are going to design is about 100 meters wide from east to west and 210 meters long from north to south. Before the demolition, there were 6 hutongs and 152 families living here. The life scene of "ceiling, fish tank, and pomegranate tree, people, fat dogs, and fat girl" will now be replaced by a group of modern multi-storey office buildings.

Respect and Seek

Entering 21st century, the speed of Beijing city renewal speeds up obviously. Feeling the gradual fading of the old city style, designers always hope to leave some memories, to retain some respect for the past life. In order to respect the gate tower of Desheng Gate, academician Cui (Kai Cui) made a diagonal street in the plot when designing the architecture, directly facing Desheng Gate. The buildings on both sides of the diagonal street irregularly enclosed several groups of courtyards, forming new hutongs and courtyard texture. This approach has set the tone for the landscape design, and the construction of more human feelings rely on the landscape design to achieve. How to continue this idea, and how to retain the memory of the city? We began the search.

场地内有什么?

寻找首先源自内部，场地内有2棵高大繁茂的古槐，一棵在街边，一棵在院内，都是原址保留。院内保留的古槐树下，还意外发现了一处砖砌的地窖，我们把这些旧有痕迹当作宝贝一样保留下来，地窖口加固后用大石板扣住，上压陶缸，既有置缸养鱼的意境，地窖遗迹也能得以保存。另外，现场还挖到了一批老城砖，大约30多块，这些老砖被散置在内院草坪中，有意识地加以强化。

除了这些可见的痕迹，我们还想到了场地内最有故事的胡同，虽然原有胡同的痕迹没有了，但是它们的名字能否留存在新的城市空间中呢？冰窖后巷、德外西后街、西马家胡同、西德丰胡同、敞风胡同、富家胡同，我们把原有胡同对应到新的内部通路上，并将胡同名称雕刻在石柱上固定在街道入口处。除了名字的保留，我们还在大槐树下制作了一处铜浮雕，把地块范围内原有的胡同肌理与开发后的楼栋叠加在一起，为德胜尚城区域的变迁过程做了形象的展示。

What's in the Site?

Searching is first from the interior, and there are two tall and lush ancient locust tree in the site, one in the street, one in the courtyard, are both reserved in the original site. We accidentally found a brick cellar under the ancient locust tree reserved in the courtyard. We keep this old trace as a treasure. The cellar after the reinforcement is buckled with a large stone plate, with the pressing of pottery cylinder. Both have the artistic conception of placing the vat to keep fish, cellar ruins can also be preserved. In addition, we also dug a batch of old city bricks in the site, about 30 pieces, and these old bricks were scattered in the yard lawn, consciously to strengthen.

In addition to these visible traces, we also think of the most storied hutongs in the site. Although the traces of the original hutongs are gone, can their names be preserved in the new urban space? Bingjiao Houxiang, Dewai Xihoujie, Simajia Hutong, Xidefeng Hutong, Changfeng Hutong, Fujia Hutong, we put the original hutong corresponding to the new internal path, and carved the names of the hutong on stone pillars fixed at the street entrance. In addition to the preservation of the names, we also made a bronze relief under the big locust tree, superimposing the original hutong texture within the plot with the developed buildings, to show the transformation process of DE Sheng Shang City.

场地外能搬来什么？

我们知道，当时的城市更新远不止德胜区域，北京城的其他地方，有许多老街区老院子正处于被拆迁的过程中。于是我们把目光扩大到北京城范围内，寻找那个时间点的城市变迁记忆，找回老院落、老门楼、老屋架的真实片段，找回拆迁下来的老砖老瓦，抢救性地保存在我们的场地中。我们花了两个月的时间，骑着自行车穿街走巷，找到拆迁现场，像收货郎一样商议买卖，然后告知甲方和施工单位具体采购迁移，每收购一处，幸福之色溢于言表。

最后，有两处北京当时拆迁的房屋片段被收集到场地内，一处选用老屋架按老法搭建，用老砖砌筑一片断墙，形成室外休息区；另一处位于古槐树下，用房山产的青白石制作了柱基础，用仿古方砖铺设了房屋地面，完全按照老北京四合院的典型格局，只是没有柱子、墙体和屋顶，这些缺失将唤起对旧有居住空间的回忆。另外还有三处老门楼，一处位于地面层院落空间内，另外两处放置在两栋楼房的屋顶，三处老门楼围着古槐树彼此相望，形成了旧有的城市脉络，与新建筑穿插并置在一起，好像旧有的胡同和房屋还存在。

除了房屋构架和老门楼，我们还从老北京城中拆迁工地的各个小工头手里寻找到老砖货源，一点点拼凑到所需要的用量，然后将老胡同和院落的肌理叠加到新建筑场地上，老胡同范围内用老砖铺设，老胡同范围外的街道用灰色透水砖铺设，很明晰地标示出了老胡同的走向，带回了老街的记忆。

景观不同于建筑，我们的造园材料是富有生命的，用老北京的传统树木是唤起城市记忆的又一策略，我们在整个区域用北京的早园竹作为基调种植，每个院落采用不同品种的树木，这些树木都是四合院常见的，从而形成藤萝院、黄栌院、玉兰院、海棠院，不同的新院落空间因为老北京的树而带来了旧时胡同院落的气息。

What Can We Bring from the Outside?

We know that the urban renewal at that time was much more than the Desheng area. In other parts of Beijing, there were many old neighborhoods and old yards in the process of being demolished. Therefore, we expanded our vision to the city of Beijing, looking for the memory of the change of the city at that time point, looking for the real fragments of the old courtyard, the old gatehouse and the old roof frame, and looking for the old bricks and tiles from the demolition, and saving them in our site. It took us two months to ride bicycles through the streets and find the demolition site, negotiate the sale like a freight man, and then inform party A and the construction unit of the specific purchase and migration. Every acquisition makes our happiness shows between the lines.

Finally, two pieces of demolished house fragments in Beijing at that time were collected into the site. An old house frame is built according to the old method, and a section of wall is built with old bricks to form an outdoor rest area; the other place is under the ancient locust tree, which made the column base with the green and white stone produced by Fangshan, and laid the floor of the house with archaize square tiles. It is completely in accordance with the typical pattern of quadrangle courtyard in old Beijing , except that there are no columns, walls and roofs, which will arouse the memory of the old living space. In addition, there are three old gateways, one in the courtyard space on the ground floor, the other two on the roof of two buildings. The three old gateways are surrounded by ancient locust trees looking at each other, forming the old urban context, interspersed with the new buildings, as if the old hutongs and houses still exist.

In addition to the building frame and the old gate, we also find the supply of old bricks from the hands of the small foreman in the demolition site of old Beijing, little by little to piece together the required amount, and then superimpose the texture of the old hutong and the courtyard onto the new building site. The old hutong area is paved with old bricks, while the streets outside the old hutong area are paved with gray permeable bricks, which clearly marks the direction of the old hutong and brings back the memory of the old street.

Landscape is different from architecture. Our garden materials are full of life. Using traditional trees from old Beijing is another strategy to evoke the city's memory. We use Beijing's Zaoyuan bamboo as the base throughout the area, and each yard uses different types of trees. These trees are common in quadrangle courtyard, thus forming the wisteria courtyard, cotinus coggygria courtyard, magnolia denudata courtyard and malus spectabilis courtyard. Different new courtyard spaces bring the smell of the old hutong courtyard because of the trees in the old Beijing.

魅力之城

Charm City

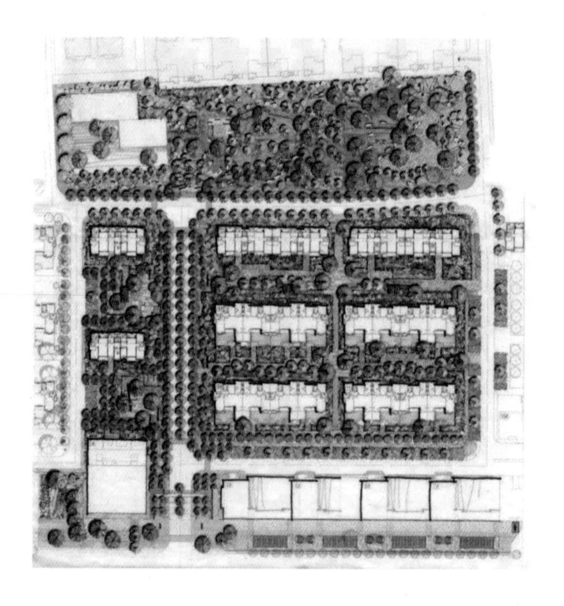

初识那片土地

冬季的北方大地一片萧瑟，我们走在沈阳城北的郊野，万科在这里将建设新的居住区，我们也将负责新区的园林景观设计。在北方冬日的寒风里，望着眼前已完成拆迁的荒野，我们要为居住在这里的人们提供一份什么样的生活？给他们呈现什么样的过往？曾经居住在这里的人们如果回来，我们能做些什么？我们带着问题寻找，而场地也试图给我们以答案。我们发现了荒草中的老磨盘，横卧的老式穿孔水泥电线杆；而场地中还有一些大榆树，它们伸向天空的枝干似乎在讲述曾经的故事。

First Acquaintance with the Land

In winter, the northern land is bleak. We were walking in the countryside to the north of Shenyang, where Vanke would build a new residential area and we would also be responsible for the landscape design of the new area. In the cold wind of winter in the north, looking at the wilderness that has been demolished in front of us, what kind of life should we provide for the people living here? What kind of past is presented to them? What can we do if the people who used to live here come back? We looked for with questions, and the site tried to give us answers. We found old grindstones in the weeds, and old perforated cement telegraph poles lying across; there are also large elm trees in the site, and their skywards branches seem to tell an old story.

家园的理想

家园，对于人们来说是栖身之所，是精神依托。城市的生长更新，常使人们感受到回不去的故乡，在乔迁新居之时难免也会有一份怅然。在城市化的过程里，那一片片的新区中，人们能找到安放心灵的家园吗？家园，是亲人的相伴、是邻里的共情；家园是院中的一棵大树、是小伙伴嬉戏的花园、是祖孙相伴而坐的廊下、是捉过蛐蛐的那段院子里的围墙、是河溪里畅游的鸭群……

“大院”生活

沈阳市是一个老工业城市，以大型国企、学校、科研院所、政府、军队为主要构成的城市，人们对于家园的概念常常是与各个大院联系起来的，并在大院里形成各自的生活记忆以及文化与情感的依托。

重塑“大院”生活是我们对于这片土地的设计理想。在“大院”的精神内涵与情境营造中，我们通过对时间与空间不同的探索与解读，从中提炼能赋予家园情感的积极元素，珍惜这里留存的土地记忆，保留场地内的原生大树、水泥电线杆、磨盘、传统工业化制作的花格。营建大院里的田园诗画，提供邻里无界交往的空间，建立童年之乐的场地，形成层层递进，外紧内松，具有归属感的“院子”空间，使之成为既有时代特性又满是童年记忆的“大院”生活家园。

Ideal of Homeland

Homeland, is a place to stay for people, is the spiritual support. The growth and renewal of the city often make people feel that they cannot go back to their hometown, and it is inevitable that they will feel a sense of loss when they move to their new home. In the process of urbanization, at that piece of new areas, people can find a place to put their hearts? Homeland, is the companion of relatives and the empathy of neighbors. Homeland is a big tree in the yard, is the garden where friends played, is the porch where grandparents and grandchildren sit, is the fence in the yard where crickets were caught, is the ducks swimming in the river···

"Courtyard" Life

Shenyang is an old industrial city, mainly composed of large state-owned enterprises, schools, research institutes, government and army. People's concept of homeland is often associated with each courtyard, and form their own living memories, cultural and emotional support in the courtyard .

Reshape the "courtyard" life and carry out our ideal design for this land. In the spiritual connotation and situation construction of "courtyard", we have explored and interpreted different time and space to extract the positive elements that can endow the homeland with emotion, cherish the retained land memory here, and keep the original trees, cement telegraph poles, millstones and traditional industrial lattice in the site. The construction of pastoral poetry and paintings in the courtyard provide a space for neighbors to interact with each other without boundaries, and establish a playground for the pleasures of childhood. A "yard" space with a sense of belonging is formed, which is progressive, tight outside and loose inside. It becomes a courtyard homeland full of childhood memories as well as the characteristics of the times.

天拖老工业区

Tiantuo Old Industrial Zone

伤感的调研

大雪过后的严寒之日，我们初次来到位于天津市区的天津拖拉机厂旧址进行调研。也许是凛冽的寒风，也许是清晨的积雪，也许是遗弃的厂房，也许是摇摆的枯树，也许是刺眼的阳光，也许是这些元素叠加构成的场景使我们始终沉浸在悲壮的情绪中不能自拔。

随着城市更新的“步步紧逼”，天拖厂区在时代更迭中战略东移，老厂址被赋予更为复合的城市功能；老厂区内大部分工业遗存被当作废品出让，只遗留六栋老厂房的“躯体”与满园大树形影相伴。走在雪中的场地里，走在空空的厂房中，我们似乎能听到曾经从厂房中传出的口号声，似乎能看见如火如荼工作的场景。但轰轰作响的车间如今早已人去楼空，热火朝天的厂区如今寂静如野，一切死一般地沉寂在城市的中央。然而，耸立云间的烟囱、茂密挺拔的林荫、散落园区的零件、革命年代的标语等似乎又正在昭示着一股新生力量的萌发。

Sad Research

At a cold day after the snow, we first came to the old site of Tianjin tractor factory located in the urban areas of Tianjin for research. Maybe it was the cold wind, maybe it was the snow in the morning, maybe it was the abandoned factory, maybe it was the swaying dead trees, maybe it was the dazzling sunshine, maybe it was the scene composed of these elements that made us immersed in the solemn and stirring emotion.

With the "pressing step by step" of urban renewal, Tiantuo factory moved eastward strategically in the changing times, and the old factory site was endowed with more complex urban functions. Most of the industrial remains in the old factory were sold as waste, leaving only the "bodies" of the six old factories and the trees in the garden. Walking in the snow in the site, walking in the empty factory, we seemed to be able to hear from the once slogan sound from the factory, seemed to be able to see the scene of working in full swing. But the noisy workshops were now empty, the bustling factory was now as quiet as the wild, and there was a dead silence in the middle of the city. However, the chimneys between the towering clouds, the thick and straight trees, the parts scattered in the park, the slogans of the revolutionary era and so on seemed to be showing the germination of a new force.

浪漫的追忆

天津拖拉机厂曾经作为新中国重工业的标志享誉中外，承载着一代人的骄傲与回忆，也见证着那个时代的光辉印记。但如今，保留的历史老厂房、现状植被、工业遗迹等结合商业开发，将被整体打造成为城市核心风貌景观区，以全新面貌融入周边城市发展。

为了更加真实地感受老一代天拖人的情感记忆，我们通过朋友介绍，采访了曾经在此工作了一辈子的老两口。在交流过程中，二老始终激情澎湃地回忆过往："这个楼是冲压车间……，这个厂房是总装车间，毛主席来过！……这个是工会，那谁和那谁是在这认识的，后来结婚了……"。交流整整持续了一个下午，我们深深感到这不仅是他们对厂区的回忆，更是对他们青春的浪漫追忆，一代天拖人根植于心的情怀难以抹去。

Romantic Memories

Tianjin tractor factory, once as a symbol of heavy industry in new China, enjoys a good reputation both at home and abroad. It carries not only the pride and memory of a generation, but also the glorious mark of that era. However, today, the preserved historic factories, existing vegetation, industrial relics, and so on are combined with commercial development, will be integrated into the urban core landscape area and integrated into the surrounding urban development with a new look.

In order to truly feel the emotional memory of the old generation of Tiantuo people, we interviewed the old couple who had worked here all their lives through the introduction of a friend. In the process of communication, the old couple always recalled the past with passion; "This building is the stamping workshop··· this workshop is the assembly workshop. Chairman Mao has been here! ··· this is the union, and the workmate met his wife here and got married··· "The exchange lasted for a whole afternoon, and we deeply felt that this was not only their memories of the factory, but also their romantic memories of their youth, and the feelings rooted in the heart of a generation of Tiantuo people could not be erased.

激情的创作

如何保留工业记忆又能赋予新的时代含义；如何利用现状资源又能建构空间秩序；如何尊重历史气质又能适应商业开发，重塑人性活力；如何满足文脉传承又能对接城市未来等一系列的问题给项目带来更多可能性。

其实我们一开始就确定了将老厂区视为生命有机体的策略，从功能、文化、生态、历史等诸多视点重新梳理场地条件，用“景观语言”形成“血脉”连接建筑之间的孤立，弥补建筑与自然之间的裂痕，重构人、建筑、自然、历史之间的和谐关系，还冰冷的工业遗址以新生的活力。

“躯体”犹在、“血脉”未存、“厂房已死”成为现实！但是面对纷繁的场地信息与历史线索，如何对厂区内保留的六栋厂房进行激活成为设计难点。我们觉得有无数条线可以抓，但是又觉得“不过瘾”。我们希望找到一个合理且具有说服力的线索而同时又不会显得刻意牵强。

“用曾经的工业流程作为线索串联全区吧！”我们提出了想法：“所有工业建筑之间都应该是串联的，生产的产品要最便捷地从一个车间传递到另外一个车间，直至成品，我们就用这个线索来设计！”“对啊，这个概念和采访的老职工对厂区生产的描述不谋而合。”一语点醒梦中人，这似乎成了最合理的答案！

很快，我们就以天拖厂区曾经的生产工艺流程为内在线索，串联各个工业厂房与室外空间，还厂房以“血脉”！打造了一个集历史展示、交通联系、休闲娱乐、艺术展陈、便民服务、生态修复的复合系统，贯穿全园，给零散分布的工业片段一个有迹可循的主线。

Passionate Creation

A series of problems, such as how can it retain the industrial memory and be given new era meaning; how can it make use of current resources and construct spatial order; how can it respect the historical temperament and adapt to the commercial development, restoring the vitality of human nature; how can it meet the cultural heritage and connect with the future of the city, bring more possibilities to the project.

In fact, from the very beginning, we determined the strategy of treating the old factory as a living organism, and rearranged the site conditions from the perspectives of function, culture, ecology, history and so on. The "landscape language" becomes the "blood line" connecting the isolation between the buildings, making up the rift between the buildings and nature, reconstructing the harmonious relationship between people, architecture, nature and history, and bringing new vitality to the cold industrial sites.

"Body" is still here, "blood" does not exist, "factory has died" become a reality! However, in the face of numerous and complicated site information and historical clues, how to activate the six reserved plants in the factory has become a design difficulty. We feel that there are countless lines to catch, but still feel "not enough." We want to find a reasonable and compelling clue that doesn't seem too far-fetched.

"Connect the whole region with old industrial processes!" We came up with the idea, "All industrial buildings should be connected in series, and the products produced should be transferred from one workshop to another in the most convenient way to the finished products. We use this clue to design!" "Yes, this concept coincides with the old employee's description about factory production." Only one word wakes up dreamers, and this seems to be the most reasonable answer!

Soon, we used the technological process of production of the Tiantuo factory as the internal clue, connecting the various industrial plants and outdoor space, returning the "blood" to plant! A complex system of historical exhibition, transportation connection, leisure and entertainment, art exhibition, convenience service and ecological restoration is built throughout the park, providing a traceable main line for the scattered industrial segments.

2013年
天房融创携手以103.2亿元竞得天拖地块，
1958年4月10日
我国第一台中马力轮式拖拉机在天津

忘却的纪念

从调研开始到创作中途，甚至直至今日，我们头脑中始终浮现着鲁迅先生的《为了忘却的纪念》，其实大家也记不清楚这篇文章里的细节内容，甚至连这个稍有拗口的题目也说不清楚究竟是想“忘却”，还是想“纪念”，但是它始终浮现！为此，我们又查了一下原文和注解，原来当时他在悲愤之余想忘却，其实是为了更好地战斗。正如我们面对历史项目一样，一方面绝不能忘记，但也不应只停留在怀念之中！

地面铜雕徐徐展开在迎来送往间诉说厂史，红砖厂房与纯净水面在交相呼应中沐浴余晖，开阔草坪中的拖拉机与远处的铁牛雕塑遥相对视，静静地诉说着曾经的回忆与骄傲。

我们为了忘却的纪念！其实我们从未忘却！

Forgotten Remembrance

From the beginning of research to the middle of creation, even to this day, Lu Xun's "For the Memory of Forgetting" always comes to our mind. In fact, we also do not remember what is said in this article, even this awkward-sounding topic also can not say clearly is to want to "forget", or want to "commemorate", but it is always emerging! For this reason, we later looked up the original text and notes, and it turned out that why he wanted to forget in grief and anger is actually for a better fight. As we face the history of the project, on one hand, we must not forget, but also should not stay in memory!

The ground bronze carvings spread out and tell the history of the factory in welcome and delivery, the red brick workshop and the pure water surface are bathed in the afterglow in response to each other, the tractor in the open lawn looks at the iron ox sculpture in the distance, quietly telling the memory and pride of the past.

We actually never forget the remembrance that we want to forget!

自然篇

Nature Imagery

自然

自然，是自然而然的客观存在。它无时无刻不影响着我们。我们赖以生存的自然承载着海洋陆地、山川河流，也承载着生长在其中的万物生灵。在没有人类活动的地方，自然以原生态的方式延续着自身的逻辑，我们要好好保护自然，尽量不要干扰它。

如果我们的设计要延伸到人类活动的地方，我们就要充分尊重自然，尽可能地减少对自然的干扰。即使在人类聚居的城市和乡村，我们也要将既有环境中的自然痕迹充分保留，我们还要对因人类生产生活而导致的生态破坏进行有效的修复，我们要以水为友、保护林木、回归田园，让自然恢复自我平衡的能力。

我们也知道，自然是随着人类活动的变迁而发生变化的，人类文明和历史也会慢慢地成为自然的一部分，在这样的历史环境里，我们的设计既要尊重自然，也要尊重历史，保护好人类文明的痕迹，并使历史文化得以充分地展示，让现代人的活动融入历史、融入自然，也成为自然的一部分。

Nature Imagery

Nature is an objective existence in a long history, and during this period, it affects human all the time. We live in the environment which carries oceans, lands, mountains, rivers and other natural creatures, it is our duty to protect our natural homeland by keeping it away from excessive artificial interference.

If the design has to extend to the place with human activities, we should fully respect nature law and minimize human disruption as much as possible. Even though the design project located in human populated cities or villages, we need to fully preserve the existing natural traces in the environment while restoring the damaged ecosystem. Designers protect water, forest, and village, and help nature regaining its equilibrant condition.

As we know, nature changes as human activities change, and on the other side, human civilization and history is becoming a part of the nature. Under such historical background, our design should respect both nature and history, protect the traces of human civilization, and fully display our historical culture, so that the modern people can integrate themselves into the history and nature, and then also immerse themselves into the nature as we all did.

第三章

自然的表现

Chapter Three

Expression of Nature

一　回归田园
Return to Pastoral

无锡鸿山遗址公园
Wuxi Hongshan Historic Site Park

田园中的遗址

鸿山遗址位于无锡新区鸿山镇东南部，北倚泰伯河，南抵新开河，西靠鸿山路，东接漕湖，与苏州相望，范围达5平方公里。遗址分布区地势平坦，田园密集，水网密布，河流湖泊众多，是典型的江南水乡自然环境，田园湿地特色明显。博物馆景区以鸿山墓群中规模最大的土墩墓-丘承墩为中心布置。

保护田园湿地

“上善若水，水善利万物而不争，处众人之所恶，故几于道”。只有充分尊重遗址本体和周边环境，保留原有地形地貌等自然环境和农田、湿地、水系肌理形态，减少人工雕琢的痕迹，才是最大限度地尊重遗址及其背景环境。通过利用原有水网，将河道清理疏通，推进湿地景观的保护，通过修理田埂，保护农田格局和农田肌理。

田园肌理布局

空间的布局方式来源于保护好田园湿地肌理，以田为底，以水为界，以路为脉，将场地梳理成两个不同类型的空间。水系以内是核心保护区域，要保护丘承墩的原生地形地貌，保护周边的农田肌理，保护农田的耕作方式，一切呈现出原始的状态。水系以外是提升区域，在充分尊重现状肌理的基础上，尽可能少地设置满足基本功能的服务设施和休闲场所，整体景观风貌以展现鸿山遗址浓重的文化历史感和沉郁的苍凉氛围为主。

Historic Site in the Countryside

Hongshan site is located in the southeast of Hongshan Town, Wuxi new district, with the Taibo River to the north, Xinkai River to the south, Hongshan Road to the west, Cao Lake to the east, and facing Suzhou, covering an area of 5 square kilometers.The geographical area of the site has flat terrain, dense countryside, dense water network and numerous rivers and lakes. It is a typical natural environment of a water town in the south of the Yangtze River and its pastoral wetland characteristics are obvious. The scenic area of the museum is arranged around Qiuchengdun, the largest mound tomb in Hongshan tomb group.

Protecting Garden Wetlands

"Best is like water, water is good for all things without disputing, and stay in a place that people do not like, so it is the closest to 'Tao' ". Only by fully respecting the site itself and its surrounding environment, preserving the natural environment such as the original topography and landform, as well as the texture of farmland, wetland and water system, and reducing the traces of artificial carving, can we respect the site and its background environment to the maximum extent. Through the use of the original water network, the river will be cleared and dredged, promoting the protection of wetland landscape through repairing the ridge, and protecting the farmland pattern and texture.

Countryside Texture Layout

The layout of the space comes from the protection of the countryside wetland texture, with the field as the base, the water as the boundary and the road as the vein, combing the site to form two different types of space.Within the water system is the core protection area, so it is necessary to protect the original topography and landform of Qiuchengdun, protect the surrounding farmland texture, protect the farming methods of farmland, all presenting the original state. Outside the water system is the promotion area. On the basis of fully respecting the current situation and texture, the service facilities and leisure places that meet the basic functions are set up gently. The overall landscape is dominated by the strong sense of culture and history of Hongshan ruins and the gloomy and desolate atmosphere.

乡土材料运用

灰砖、荒石是江南民居建筑砌筑的主要材料，其所表达的气质和意象正好符合遗址背景环境的氛围。于是，荒石碎拼便成了主要的铺地方式，透过宽宽的缝隙，野花野草沿缝生长，形成自然古朴的氛围。田间绿化也主要采用当地的农作物，根据时令种植，呈现出浓浓的乡土气息。

Application of Native Materials

Grey bricks and waste stones are the main materials used in the construction of residential buildings in the south of the Yangtze River. The temperament and image expressed by them are just in line with the background environment of the site. Then, the waste stone fragments became the main way to pave the ground, through the wide gap, wild flowers and weeds along the seam growth, forming a natural and simple atmosphere. The field afforestation also mainly uses the local crop, according to the seasonal planting, presenting the thick local flavor.

二 生态修复

Ecological Restoration

永城日月湖

Riyue Lake in Yongcheng

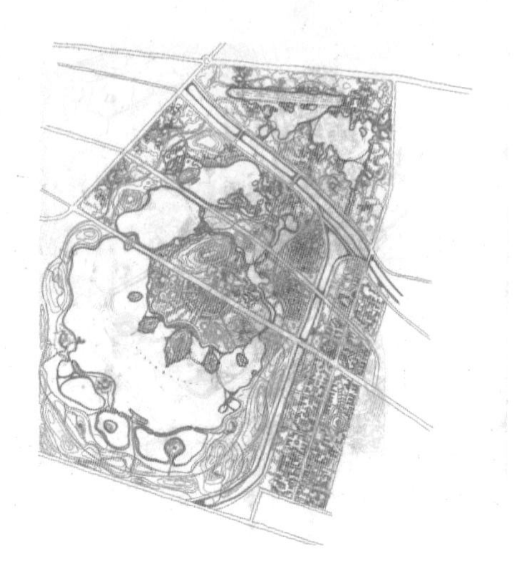

“那片麦田”

一片片嫩绿的小麦与枯黄的苇草随风轻拂，远处平静的水面上点缀着一栋栋民房，干硬的泥土夹杂着煤矸石、青砖和瓦片，突然的脚步声惊起了水边发呆的野鸭。还记得那是2011年的晚秋，我们初次踏勘现场的景象，呈现在眼前的是宛若油画般的中原美景。

进一步深入场地，地势渐低，地面塌陷，出现了一个个坑塘，破损残存的村庄道路、房屋东倒西歪地淹没在水面中，满目疮痍的土地像一位受难的老者被肢解着。工作人员介绍，基地因早期煤炭的开采日渐塌陷，大量煤矿废水、粉煤灰及露天堆砌的煤矸石对环境已造成恶劣的影响。我们爬上土坡静静地注视着眼前的一片水泽，这个尺度巨大的18平方公里的煤矿塌陷区治理项目就这样在脚下开始了。

"The Wheat Field"

Flecks of verdant wheat and withered reeds gently brushed in the wind, the peaceful waters in the distance were dotted with houses, the dry soil was mixed with gangue, black bricks and tiles, and the sound of strange footsteps startled the wild ducks at the water's edge. We still remember that it was the late autumn of 2011, when we visited the site for the first time, the beautiful scenery of central plains presented like oil paintings before our eyes.

Further into the site, the terrain gradually lowered, the ground collapsed, and a series of pits appeared. The damaged and remaining village roads and houses were submerged in the water. The scarred land looks like a suffering old man being dismembered.Staff introduced that the site is gradually collapsing because of coal mining in the early time, and a large amount of coal mine waste water, fly ash and coal gangue piled up in the open air have caused bad influence on the environment. As we climbed up the soil slope, we watched silently the pool of water in front of us. The massive 18-square-kilometer administration project of mine subsidence area began beneath us.

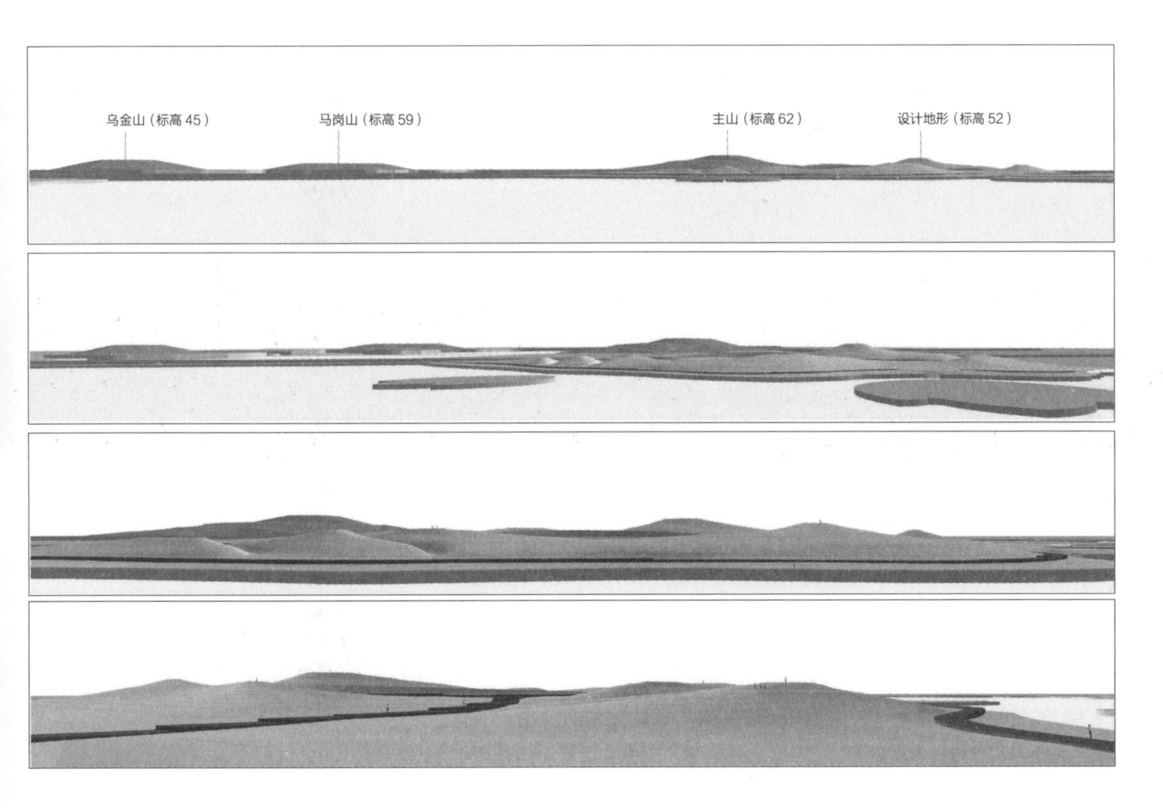
乌金山（标高 45）
马岗山（标高 59）
主山（标高 62）
设计地形（标高 52）

“黑白转绿”

永城，全国六大无烟煤基地和河南省最大的煤化工基地，国家百强县市，有着“面粉城”和“煤炭城”之称。煤矿塌陷区治理这个项目伴随着城市的转型，居民对公园绿地的需求日益旺盛，因塌陷区面积大，人地矛盾突出，治理方法和措施上存在建设周期长、资金投入大、市场开发程度弱等问题，关于区域的定位因此变得尤为重要。

我们提出“心象自然”的理念，强调让自然做功，生态先行，确定大山大水的空间格局，培育生态绿地重构场地的生态平衡，采用“生态提升品质、功能提升活力、艺术提升魅力、商业提升价值”的规划设计理念，循序渐进地推进工作。在规划设计过程中，我们渐渐体会到心中的那片绿色对于城市、对于土地的重要，脑海中总是浮现骑着自行车的小孩追着我们问，这里什么时候能建成一个大公园?

因为这份隐隐的责任，我们使出十八般武艺，利用沉陷区所形成的自然水面，通过塑模型的搭建并采用GIS及土方平衡软件反复推敲，调整设计地形，确定山体的体量、高度、坡度、坡向，使原有残破的场地形成延绵起伏、大山大水的艺术空间。通过软件模拟，合理组织挖方与填方工程，减少了大量土方工程费用。

运用“理水堆山”的造园手法，根据场地现有的坑塘确定水形和山形，结合永城特色的黑白经济，模仿“麦穗”和“乌金”，形成特色岛和湿地岛屿的形象，重点研究水陆边界的特色及山水相依的关系，增加景观层次，形成内湖、外湖、岛屿、湿地、堤、塔、亭、桥错落有致、分布有序、丰富而有特色的山水空间格局。因其最终的场地结构形如太阳，水体围绕周边，宛如一弯明月，故名“日月湖”。

"Black and White Turns Green"

Yongcheng is one of the six largest anthracite coal bases in China and the largest coal chemical base in Henan Province, national top 100 counties and cities, well-known as "flour city" and "coal city" . The administration project of mine subsidence area has been accompanied by the transformation of the city, and residents have a growing need for parks and green spaces. Due to the large subsidence area, prominent contradiction between people and land, long construction period, large capital investment, weak market development and other problems in governance methods and measures, the positioning of the region has become particularly important.

We put forward the concept of "nature in mind", emphasizing that let nature do the work and ecology take the lead. Determine the spatial pattern of large mountains and large rivers, cultivate ecological green space to reconstruct the ecological balance of the site, and adopt the planning and design concept of "ecology improves the quality, function enhances the vitality, art enhances the charm and commerce enhances the value". Gradually promote the work, gradually realize that the importance of the green in the heart for the city and the land. We always think of in the mind that kids on bikes chasing us and asking, when will there be a big park?

Because of this hidden responsibility, we use all skills we have. By using the natural water surface formed by the subsidence area, through the construction of Su model and repeated elaboration by GIS and earthwork balance software, the terrain design was adjusted. Determine the volume, height, slope, slope direction of the mountain, so that the original wrecked site forms a rolling art space with large mountains and large water. Through software simulation, reasonable organization of excavation and filling engineering reduce a lot of earthwork costs.

By using the landscape technique of "combing water and piling mountains", the shape of water and mountain was determined according to the existing pit and pond on the site, and the black and white economy with Yongcheng characteristics was combined to imitate "ear of wheat" and "Wu Jin" to form the image of characteristic island and wetland island. The research focuses on the characteristics of the boundary between land and water and the relationship between mountains and rivers, and increases the level of landscape, so as to form a well-arranged, well-distributed, rich and distinctive landscape spatial pattern of internal and external lakes, islands, wetlands, embankments, towers, pavilions and bridges. Because of its final site structure like the sun, water around the surrounding like a curved moon, hence it is named as "Riyue Lake".

“八年守望”

从规划到具体的详细设计，如今已有8年的时间，整个日月湖形态逐步得以显现，形成北部以人文景观为主、中部以艺术景观为主、南部以自然景观为主的日月湖生态景区。我们喜欢通过google影像图一月月一年年看着日月湖的变化，自我陶醉地欣赏着设计在大地上的印记。每次去现场看到熙熙攘攘的人群，在草坡上打滚嬉戏的孩童，在山坡上放风筝游乐的老者，心中无比的自豪。

一个原本毫无生机、对城市有负面影响的塌陷区，焕发了新的生命，实现了资源型城市的生态转型，日月湖景区的建设成为永城和谐发展的催化剂。通过“去黑存绿”的生态改造，日月湖景区成为城市的绿肺，如同杭州的西湖，成为城市的代言。目前，日月湖景区已被授予国土资源部环境治理示范工程、国家级水利风景区、国家级文化旅游产业示范区及河南省低碳旅游示范区，并筹划申报国家4A级景区。

回顾项目的历程，作为设计师最大的欣慰是通过设计使生态环境得以恢复，成就了心中的那份美好。还记得接送我们的司机师傅聊起日月湖，说因为没机会带孩子看大海，小孩看到湖面时兴奋地问他“这是不是大海？”正如孩童心中的那片海，“心象自然”这个主题或将是我们永恒的追求，用自然反映人性的真实，用内心体悟自然的美好。

"Eight Years Watch"

From planning to detailed design, it has been eight years now. The whole shape of the Riyue Lake gradually emerged, forming an ecological scenic spot of the Riyue Lake, cultural landscape in the north, art landscape in the middle and natural landscape in the south. We love to see the changes of the Riyue Lake through Google images month by month and year by year, enjoying the imprint of design on the earth. Every time to see the scene of the bustling crowd, the children rolling on the grassy slope and having fun, the elderly playing kite in the hillside and amusing, we are very proud the heart.

A subsidence area that had no vitality but had negative impact on the city has gained new life and realized the ecological transformation of the resource-based city. The construction of Riyue Lake scenic spot has become the catalyst for the harmonious development of Yongcheng. Through the ecological transformation of "eliminating black and preserving green", it has become the green lung of the city, just like the West Lake in Hangzhou, which has become the representative of the city. At present, it has been awarded the environmental management demonstration project of the ministry of land and resources, national water conservancy scenic spot, national cultural tourism industry demonstration area and Henan province low-carbon tourism demonstration area, and plans to apply for the national 4A scenic spot.

Reviewing the course of the project, the greatest comfort as a designer is that the environment and ecology can be restored through design and the beauty in the heart has been achieved. We still remembered the words that our driver who picked up us said about the Riyue Lake, because there is no chance to take the child to see the sea, the child excitedly asked him when he saw the lake "is this the sea?" Just like the sea in the heart of children, the theme of "nature in mind" may be our eternal pursuit, reflecting the truth of human nature with nature and realizing the beauty of nature with heart.

三　与水为友
Make Friends with Water

厦门杏林湾湿地公园
Xinglinwan Wetland Park in Xiamen

生态优先初识共鸣

2015年初冬，我们应待建业主的邀请前往厦门现场调研杏林湾湿地公园。这里正是2007年第六届中国（厦门）国际园林花卉博览会的举办地，水面开阔，美景如画，生机盎然。这里还需要设计吗？伴随内心的疑问，我们来到了真正的现场，杏林湾上游水口后溪水闸口处，仅仅一桥之隔，景色却完全不同。这里是一片荒地，水岸陡峭、杂草丛生，水质暗绿、黑臭严重。面对环境问题如此严峻的现场，生态优先治理，景观“顺势而为”成了最直接的想法，与业主现场当面沟通后，得到共鸣首肯，合作也由此拉开序幕。

First Resonance on Ecology First

In the early winter of 2015, we went to the site in Xiamen to investigate Xinglinwan Wetland Park at the invitation of the owner. It is the site of the 6th China (Xiamen) International Garden Expo in 2007, where the water is wide, picturesque and full of life. Does it still need to be designed? With inner doubts, we came to the real scene. At the mouth of a stream gate behind Xinglinwan upstream, just across the bridge, the scenery is completely different. It was a wasteland, with steep water bank, overgrown weeds, and the water in dark green and black. In the face of such a severe site, making ecological priority management and landscape"go with the flow" has become the most direct idea. After face-to-face communication with the site owners, we get a sympathetic approval, and cooperation also kicked off from this.

梯田治水与微循环

随着调研的深入，现场问题也越来越清晰，要想美化水岸、改善黑臭水体，只有让岸线形成缓坡，让水流动起来才能真正解决问题，于是“梯田治水+微循环”成了最直接有效的策略，这与业主再次一拍即合。梯田治水与微循环是因地制宜按照雨水收集、雨水回渗、水质净化、黑臭治理、营造水景等五个递进层次，将表流湿地、坑塘湿地、上行湿地、生物填料、推流富氧机等借助自然重力流的力量，打造成梯田式微循环的水质净化系统，实现水质自然净化。

精益求精漫长配合

方案顺利通过，但真正实施却困难重重。一方面是竖向高差难题，现状闸口水位比梯田顶部低10米左右，直接就近取水简单方便，但需要提升泵，做法不生态，后期运维费用大。要解决此问题就只能上游取水，而要寻找到与梯田顶部高差相同的区域，须在上游几公里之外取水，距离太长同样不经济。为了解决此问题，方案将4~5级梯田一分为二处理，高处两级以雨水和“太阳能水泵”抽水为水源，低处几级采用河道重力流作为水源，由此形成多级水源保障的自然生态式梯田治水系统。另一方面是湖底淤泥难题，杏林湾的淤泥是整体性问题，无法局部单独解决，方案只能用装满种植土的麻袋放入湖中，作为挡土设施，然后配置部分种植土，种植水生植物。这两个关键难题的解决，确保了梯田湿地的自然性和景观性。

Terrace Water Control and Microcirculation

As the investigation goes deeper, the problems in the site become clearer and clearer. To beautify the water bank and improve the black and smelly water, let the shoreline gently slope up and let the water flow up can really solve the problem, so "terrace water control + micro-circulation" became the most direct and effective strategy, which hit it off with the owner again. Terrace water control and microcirculation adjust measures to local conditions on the following five progressive levels: rainwater collection, rainwater infiltration, water quality purification, black smelly governance, and waterscape construction. By taking advantage of the force of natural gravity flow, such measures as surface flow wetland, pond wetland, ascending wetland, biological packing, push flow oxygen enrichment machine and so on, the water quality purification system of terrace micro-circulation is built and realized the natural purification of water quality.

Continuous Improvement and Long Cooperation

The plan passed smoothly, but implementation was difficult. On one hand, there is the problem of vertical height difference. The current water level at the gate is about 10m lower than that at the top of the terrace, so it is simple and convenient to take water directly nearby. But it needs to lift the pump and is not ecological, with high operation and maintenance costs. The only way to solve this problem is to go upstream, and to find an area with the same elevation difference as the top of the terrace, which has to go a few kilometers towards upstream and is also uneconomical in such a long distance. In order to solve this problem, the scheme divided the 4-5 stage terrace into two parts, with rainwater and "solar water pump" pumping as the water source for the upper two stages and river gravity flow as the water source for the lower several stages, thus forming a natural ecological terrace water control system with multi-stage water source guarantee. On the other hand, it is the silt problem at the bottom of the lake. The silt in Xinglinwan is an overall problem that cannot be solved locally. The plan is to use sacks filled with planting soil to put into the lake as a retaining device, and then allocate some of the planting soil to plant aquatic plants. The solution of these two key problems ensures the nature and landscape of terraced wetland.

齐心协力实现初心

通过将近一年半的施工配合、不断探讨、修改、施工，业主、设计、施工多方齐心协力，追求自然、追求卓越，从过程中的整体效果来看，当初选择“生态优先、梯田治水”的策略是完全正确的。在通过生态梯田的打造实现水质净化的同时，也很大程度上解决了内湾水动力不足导致的水体黑臭现象。同时，通过自然重力流的方式也很好地解决了梯田补水问题，经济上可持续，生态上可示范，景观上可体验，真正实现了自然积存、自然渗透、自然净化的生态海绵理念。

Work Together to Achieve the Original Goal

Through nearly a year and a half of construction coordination, continuous discussion, modification and construction, the owner, designer and construction work together in pursuit of nature and excellence. Judging from the overall effect of the process, the strategy of "ecological priority and terrace water control" was completely correct. Through the construction of ecological terraces, the water quality is purified, and the black and smelly phenomenon caused by the lack of water power in inner bay is solved to a large extent. At the same time, natural gravity flow is also a good way to solve the terrace water replenishment problem, sustainable in economic, ecological in demonstration, experiential in landscape, truly realize the ecological sponge concept in natural accumulation, natural penetration, natural purification.

南宁园博园雨水花园

Rain Garden in Nanning Garden Expo

坝外水境

作为南宁园博园八景之一的芦草叠塘位于场地的西北侧，防洪堤之外，毗邻八尺江的一块滨江区域，现状为连片鱼塘及农用地，场地内散落多年生长的大树及花灌木，环境郊野自然。该区作为园博园外围与城市水生态环境的协调区，既要恢复滨水生境，又要在非汛期保持坑塘内的水量和水质，形成供人游赏的可持续景观。低影响开发的生态湿地系统、自我循环的生物链层级、水质净化的科普示范、具有艺术气质的独特景观，逐渐成为我们所追求的目标。

Water Environment out of Dam

As one of the eight scenes of Nanning Garden Expo, interlacing reeds and grass pond is located in the northwest of the site, outside the flood bank, adjacent to a riverside area of Bachi River. The current situation used to be a continuous fish pond and agricultural land, with perennial trees, flowers and shrubs scattered in the site and natural countryside. As the coordination area between the periphery of the expo garden and the urban water ecological environment, this area should not only restore the waterfront habitat, but also maintain the water quantity and water quality in the pit during the non-flood season, so as to form a sustainable landscape for people to visit. The ecological wetland system with low-impact development, the biological chain hierarchy with self-circulation, the demonstration of water quality purification, the unique landscape with artistic temperament, have gradually become our pursuit of the goal.

低影响雨水花园

利用现状闲置的19处水塘，分层级打造湿地雨水花园，外围的25000平方米水塘低影响开发，水量靠自然降水补给，形成干枯两季的季节性景观；核心5000平方米打造精细雨水花园，以江水提灌作为精细水花园的水源补给，配合多层次水生植物打造精美的核心景观。这样有侧重分层次的水体设计，符合投资少、见效快、低维护的经济原则。

Low Impact Rain Garden

Make use of the 19 idle reservoirs to create a wetland rain garden in different levels. The surrounding 25000 square meters reservoir will be developed with low impact. The water volume will be replenished by natural precipitation and form a seasonal landscape of two dry seasons. An elaborate rain garden is created in the core 5000 square meters, and river irrigation is used as the water supply for elaborate rain garden, with multi-level aquatic plants to create a beautiful core landscape.This layered water design is in line with the economic principle of less investment, quick results and low maintenance.

循环净化江水

为保持雨水花园中的水“从江中来到江中去”的全过程，设计借助地形变化，形成不同标高的叠塘，从而形成一套完整的水循环体系。具体包括：江水提升、水渠输水、水景跌落、地面水渠、自然叠塘、植物净化、汇入核心塘、排入八尺江。水循环为自然循环，利用场地标高形成溢流系统。根据当地气候，水体冬季不泄空。如遇干旱，可利用手提式农用泵进行临时抽水补水，以丰富雨水自然存蓄的坑塘景观。

丰富的水表情

参考当地传统水渠做法，用锈板做成新的水渠，从八尺江中提升江水注入其中，游人可沿着水渠边的栈道观察水的提升和疏导过程； 利用水渠和地面高差，形成人工叠瀑，跌入广场地面水池中，形成动态水景；水从锈板的人工水池里面流出来，顺着地面田埂路边的石渠缓缓流淌，游人边走边能听见潺潺的水声；地面蜿蜒的流水注入第一个水塘，水塘绿岛交错，水形成不同的动线，奔流向各个方向，时而分流、时而交汇。 与此同时，水在流动的过程中得到了充分的暴氧净化；水塘之间的高差形成了两组小跌水 ，进入不同池塘，进行植物净化，每个池塘植物有所不同，研究净化能力的同时进行科普展示，栈桥穿越其中，人走在上面有种凌波微步的意境；水汇集到核心塘，已经是比较清澈了，它变得平静下来，这里有千姿百态的水生植物，荷花、睡莲等，组成了精致的水花园；多余的塘水溢流进周围的池塘，和不期而遇的雨水汇集在一起，加之芦苇、菖蒲、花叶芦竹等多样的水生植物一起形成了半干的湿地草塘；水最后汇集到江边，沿地形、滩涂、洼地形成鸟类栖息区域，吸引水鸟前来。

Circulating to Purify the River Water

In order to maintain the whole process of water "coming from the river and to the river" in the rain garden, with the help of terrain changes, the stacked ponds with different elevations are formed to form a complete water circulation system. It specifically includes: river water lifting, canal water conveyance, waterscape falling, canal water on the ground surface, natural stacking pond, plant purification, into the core pond and into the Bachi River. The water circulation is a natural circulation, and the site elevation is used to form an overflow system. According to the local climate, the water does not leak in winter. In case of drought, portable agricultural pumps can be used to temporarily pump water to enrich the natural storage of rainwater pond landscape.

Rich Waterscape Expression

Referring to the local traditional practice of water channels, new water channels are made with rust plates to lift the water from the Bachi River and inject it into it. Visitors can observe the process of water lifting and channeling along the plank road beside the water channels. Using the height difference between the canal and the ground, the artificial cascade is formed and falls into the ground pool of the square to form a dynamic waterscape. Water flows out from the rust plate of the artificial pool, along the stone canal beside roadside of the ridge on the ground, and visitors when walking can hear the murmur of the water. The first pond is fed by the meandering water on the ground, which is interlaced with green islands. The water forms different dynamic lines, flowing in all directions, sometimes diverging and sometimes converging. At the same time, in the process of the flow, the water is obtained a sufficient oxygen purification; the difference in height between the ponds resulted in two sets of small drops that went into different ponds for plant purification. Plants in each fishpond are different, researching purification ability at the same time for scientific demonstration, and trestle does alternately through them, which has a sense of lingboweibu artistic conception when people walk above; Water collected into the core pond, has been relatively clear, and it becomes calm. There are a variety of aquatic plants, lotus, water lilies, forming a delicate water garden; the excess pond water overflowed into the surrounding pond, gathered together with the unexpected rain, and together with a variety of aquatic plants, such as reed, calamus and arundo donax var. versicolor, the semi-dried wetland grass pond is formed. Finally, the water collected to the river, along with the terrain, beaches, low-lying lands to form bird habitat area, attracting water birds to come.

艺术田园

以荷花、睡莲、浮萍等水生植物为主题的核心水塘周围，在一片色彩热烈的粉黛乱子草丛中，仿佛生长出深绿浅绿的“竹子”，上面又好像飘着“一片云”。这是一组特制的云形亭廊，提供遮荫措施的同时形成场地独特气质。镜面的吊顶反射着人和环境相融合的光影，魔幻而神奇，仿佛置身世外桃源。

芦草叠塘西南角有一株160年树龄的“芒果王”，成为当之无愧的镇园之宝，无论什么样刻意设计的人工环境都难以融进它的气场，于是只在树下围绕广场散置卧石，让游人在树荫下乘凉、静思。夕阳西下，江边白鹭掠过，刹那间会惊现“落霞与孤鹜齐飞”的诗情画意。还有观鸟亭、水中栈道、芳草花境等景致，设计中尽可能统一语言，田园野趣中透出艺术氛围，让人徜徉其中，流连忘返。

Art Field

At the surrounding of the core pond, it used lotus, water lily, duckweed and other aquatic plants as the theme. In an intensely colored grass cluster of Muhlenbergia capillaris (Lam.) Trin., it is as if the growth of "bamboo" in dark green and light green, and as if floating "a cloud" above. This is a set of specially designed cloud-shaped pavilion corridors, providing shade measures while forming the unique temperament of the site. Mirrored suspended ceiling reflects the light and shadow that people and environment blend together, magical and miraculous, as if in a paradise.

There is a 160-year-old "mango king" tree in the southwest corner of interlacing reeds and grass pond, which has become the well-deserved treasure of the garden. No matter what kind of deliberately designed artificial environment is difficult to integrate into its aura, so we only place lying stones around the square under the trees dispersedly, let visitors enjoy the cool and quiet thinking in the tree shade. When the sun is setting in the west, egrets fly by the riverside, and suddenly the poetic and pictorial meaning of "sunset clouds and lonely wild ducks fly together" will appear. There are also bird-watching pavilions, waterside walkways, fragrant grass and flowers, etc. Unify the language as much as possible in the design, and the artistic atmosphere permeates through the rustic charm in the countryside, let people wander in it and enjoy themselves so much as to forget to leave.

多目标景观再造

由闲置的鱼塘变成生态的雨水花园，我们探索了低影响的介入、雨水的利用、江水的循环、植物的净化，也尝试将水处理的过程转化为动态的景观场景，使人在科普展示中体会到景观的魅力，体会到功能、生态和艺术相结合的景观氛围。对原有环境的尊重、场地标高的研究、水循环和净化措施的实践、系列水景的营造、原生树木的保存、艺术手法的运用，使得我们逐渐实现了多目标协同的景观再造。

Multi-objective Landscape Reconstruction

From idle fish ponds to ecological rain gardens, we explored low-impact intervention, rainwater utilization, river water circulation and plant purification, and also tried to transform the process of water treatment into dynamic landscape scenes, so that people could experience the charm of landscape and the landscape atmosphere combining function, ecology and art in the science exhibition. The respect for the original environment, the study of site elevation, the practice of water circulation and purification measures, the construction of a series of waterscapes, the preservation of native trees, and the application of artistic techniques have gradually enabled us to achieve multi-objective collaborative landscape reconstruction.

四　保护原生
Original Protection

北京外国语大学
Beijing Foreign Studies University

初始状态

几十棵枝繁叶茂的参天大树、密不透风的植物群落、繁复叠加的材料肌理构成了北京外国语大学东校区的景观初始状态，时间的痕迹累积在场地之上，形成了丰富的视觉体验，具有一种“故园”的厚重感与场所魅力，但封闭的围合也造成了功能上的不便与情趣的缺失，然而这就是它“自然”的初始状态。

Initial State

The east campus of Beijing Foreign Studies University is originally characterized in landscape status by dozens of towering trees in full bloom, impermeable plant communities and overlaid materials texture. The traces of time accumulate on the site, forming a rich visual experience, with a gravity sense of "hometown" and charm of the site. However, closed enclosure also causes inconvenience in function and lack of interest, which is its "natural" initial state.

原生与破题

设计之初，“原生的自然”与“懵懂的心象”不期而遇；轴线、穿插、弧线、冲击力、借景、“先抑后扬”、现代、后现代等景观元素都被组织在一张草图上。然而，过多的设计改造缺少了场所的原有气质，缺少了场地特有的设计感。

这块地可不可能是一个以保护原生树为起点，以空间弹性为逻辑的“匀质空间”？就像一副抽象画那样呢?

基于原生树保护的“匀质空间”“破题心象”使我们为之一震，发自“自然”的“心象”确实才是真正的“设计感”与合理性，即使是十几年回头反思总结，找到这样的感觉依然使我们“心潮澎湃”。而后，空间结构一气呵成！后来的施工验证了这种设计手法不但从空间规划上具有完整性与合理性，更为后期植物保留、路径调整预留了巨大的灵活性与可能性。

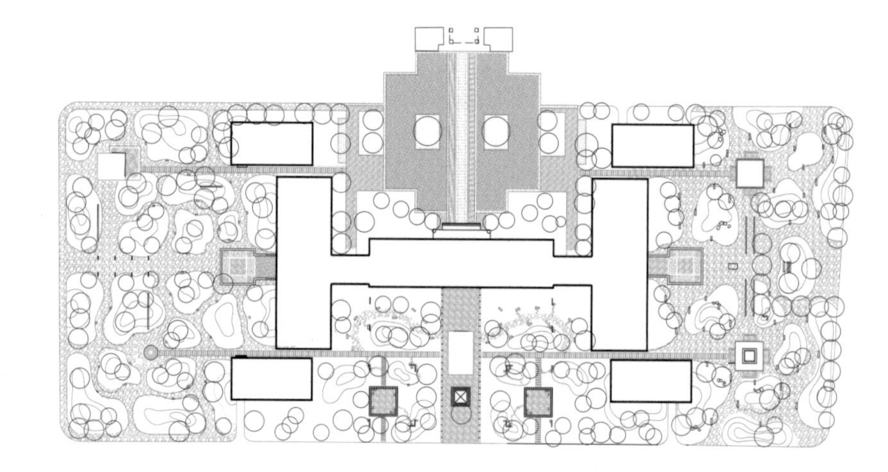

Original and Problem Solving

At the beginning of the design, "original nature" and "ignorant mind" happened to meet by chance; axis, interpenetration, arc, impact, view borrowing, "first suppress, then uplift", modern, post-modern and other landscape elements are organized on a sketch. However, excessive design and transformation are lack of the original temperament of the site, and lack of the site's unique sense of design.

Is it possible that this land is a homogenous space with the logic of spatial elasticity, starting from the protection of native trees? Like an abstract painting?

The "homogeneous space" and "solving mental image" based on the original tree protection make us shocked. The "mental image" from "nature" is really the "design sense" and rationality. Even after more than ten years of reflection and summary, finding such a feeling still makes us "surging of emotions". Then, the space structure was designed from beginning to end without stopping! Later construction verified that this design method not only had integrity and rationality in spatial planning, but also reserved great flexibility and possibility in plant reservation and path adjustment.

气韵原生的“学”“院”

对于场地“原生性”的认识其实也是层进式的，是逐层发觉、保留、借用、转化的过程。从初识的视觉认知到信息梳理，再到施工现场的古物拾遗，我们认识到场地的原生性具有更加多元的历史气韵。如何将“气韵”的原生性保留及传承下来成了又一个新的问题。最终，我们凸显了“学院” 概念。这里面包含了两层含义：即“学”和“院”。

“学”，好学近乎知——《礼记 · 中庸》。“学”是一种通过模仿而得到的提高，更是一种互动和交流。“院”，有垣墙者曰院。——《增韵》。“院”是中国人与天交流、与人沟通的最佳场所，是人们千百年来最为认同和追求的空间形式，反映了传统文化的精髓。因此在对北京外国语大学校园景观的改造设计中我们试图找寻“学”与“院”的关联，用独特的手法重新诠释“学”“院”的概念。

阳光下，斑块状的绿色草坡与保留的大树交相呼应，构成了匀质的交流空间，一种自然、亲切、多元的景观意向感染着每一个使用者。设计者运用这种手法不仅成功地营造出“学”的空间氛围，更是巧妙地解决了学习交流、交通流线、尊重现状等问题。同时，植入“墙”的元素，构成具有围合感的院落空间，形成具有东方空间魅力的、现代的、多层次的、含蓄的环境场所，不仅大大提升了校园的文化品位，也赋予其新的景观含义，找寻到了场所应有的记忆。

"Study" and "Courtyard" of Original Artistic Conception

The understanding of the "original nature" of the site is actually a progressive process, which is a process of discovery, retention, borrowing and transformation. From the initial visual cognition to the information combing, and then to the collection of antiquities in the construction site, we realize that the original nature of the site has a more diverse nature of history. How to preserve and pass down the original "artistic conception" has become a new problem. Finally, we highlight the concept of “study courtyard”. This contains two meanings: "study" and "courtyard".

"Study", study is close to knowledge — "Book of Rites · The Mean". "Study" is a kind of improvement through imitation, but also a kind of interaction and communication. "Courtyard", there are walls called courtyard. — “Zengyun”. "Courtyard" is the best place for Chinese people to communicate with the sky and with people. It is the space form most recognized and pursued by people for thousands of years, reflecting the essence of traditional culture. Therefore, in the reformation design of the campus landscape of Beijing Foreign Studies University, we try to find the relation between "study" and "courtyard", and reinterpret the concept of "study" and "courtyard" with unique methods.

Under the sunshine, the patches of green grassy slope and the reserved trees echo each other, forming a homogenous communication space, and a natural, kind and diversified landscape intention infects every user. Designers use this technique not only to successfully create a "study" space atmosphere, but also to skillfully solve the study communication, transportation streamline, respect the current situation and other problems. At the same time, the element of "wall" is inserted to form a courtyard space with a sense of enclosure, forming a modern, multi-level and implicit environment with oriental space charm. Not only greatly improved the cultural taste of the campus, but also gave it a new landscape meaning, to find the required memory of the place.

鸟巢原生树

Bird's Nest Native Tree

场地的原住民

2004年春天，成立不久的景观所开始了一项充满挑战的工作，与瑞士赫尔佐格和德梅隆设计事务所合作进行鸟巢的景观设计。在场地踏勘时我们发现，现场有很多原生树木，品种有毛白杨、柿树、雪松、悬铃木、刺槐、核桃、杜仲、丁香等。虽然这些都不是古树名木，但很多树木规格都较大，长势也很好。尤其是场地南侧几十棵毛白杨非常高大挺拔，自然围合出场地的边界。我们突然感觉到，这些树木才是这块场地的原住民，他们不仅维持着场地的生态平衡，还能够见证这片土地的过往和未来，于是我们提出保护原生树木的想法。

努力后的欣慰

想法提出后，得到了内部和外方合作设计师的认可，也得到了业主的支持。可是实现这个想法的过程却很艰难。国家体育场用地很有限，建筑承载的功能又十分庞杂，建筑之于场地，好比一个大胖子坐在一把小椅子上。在复杂的场地上保留树木，无疑要给自己增加很多麻烦：现场选择、逐一标记、图纸定位、设计中各专业的协调、管线综合时的避让、施工过程中的保护……如今，每当看到那些白杨树依然挺拔，微风吹过，树叶沙沙作响，仿佛倾诉着因奥运而发生的故事，我们的各种努力中伴随的烦恼便一下子随风而逝了。

Indigenous People of the Site

In the spring of 2004, the newly established landscape institute began the challenging task of collaborating with Swiss firm Herzog & de Meuron on the landscape design of the Bird's Nest. During site survey, we found that there were many native trees in the site, including populus tomentosa, persimmon tree, cedar, sycamore, locust, walnut, eucommia ulmoides and cloves. Although these are not famous ancient trees, but many of the size of the trees are large, growing well. In particular, dozens of poplars on the south side of the site are very tall and straight, naturally enclosing the boundary of the site. We suddenly felt that these trees were the original inhabitants of the site, not only maintaining the ecological balance of the site, but also witnessing the past and future of the land, so we came up with the idea of protecting the native trees.

Relief After the Effort

After the idea was put forward, it was recognized by the interior and foreign cooperation designers, and also got the support of the owners. But the process of realizing this idea was difficult. The national stadium has very limited land, and the functions of the building are very complex. The building on the site is like a fat man sitting on a small chair. In the complex site to retain trees, there is no doubt to add a lot of trouble to their own: site selection, marking one by one, drawing positioning, design coordination of various professionals, avoidance of pipeline integration, protection in the construction process, and so on…Nowadays, whenever we see those poplars are still tall and straight, the breeze blows, the leaf rustle is rustling, as if talking the story happened because of the Olympic Games, our each kind of effort that brought the trouble to be gone with the wind suddenly.

长辛店项目

Changxindian Project

原生的美

长辛店项目位于北京丰台区，是区域性持续开发的项目。十几年前，我们先是做策略性的研究，而后为业主做了一级开发的拿地规划，直至后期几个项目的设计落地，一路下来也是一种机缘。清楚记得最初调研场地的原生状态，成片茂盛的核桃树、枣树、柿子树生长在缓缓的山坡上；浅山的村庄里，乡里乡亲平静而悠闲地生活着。那时候北京六环路还没有通，那时候园博园还没有建，那时候一切是平静的、安详的、自然的，具有原生的美感。

Beauty of Native

Changxindian project is located in Fengtai District in Beijing, and is a regional sustainable development project. More than a decade ago, we first did the strategic research, and then did the land planning of A level development for the owner, until the later completed constructions of several project design, all the way down is also a kind of opportunity. We clearly remembered the original state of the original research site, where lush walnut trees, jujube trees and persimmon trees grew on the gentle hillside. In the shallow mountain village, the villagers live a quiet and leisurely life. At that time, the sixth ring road had not been open to traffic, the garden expo had not been built, everything was calm, peaceful, natural, with the original beauty.

原生的树

我们始终无法忘怀那早已远去的原生力量，因此，我们向业主提出，要想留住“乡愁”，首先应最大限度地保留、移栽现状树木，它们是场地的记录者、见证者，而且更是伴随场地继续前行的主人。这个策略很快得到业主的认同，但随之而来的问题是由于建筑规划大面积地下车库，现状树木均需要就近移植，而移植的费用将远高于重新购买苗木的价钱，而且由于施工季节处于初夏，移栽的成活率不能得到保证。那么，移还是买？业主有点犹豫不决。于是我们提出根据不同树木的情况，编制不同的保护策略，给每一棵树进行编号，逐一提出移植及养护方法，最大限度地保证移栽的成活率以及最低的移栽成本。我们的专业性给业主吃了定心丸，保护、移栽现状树木得以实现。最终，通过不断的努力，在一期范围内，原地保护树木8棵（包括两棵挂牌古树），就近移栽20余棵。多年来，每次回访项目，看到这些被保护的树木，都好像故友重逢，沙沙的树叶响声也好像在向我们致谢。

原生的物

场地内原存一座土地庙，经过文物部门的鉴定后认为，该庙虽然距今有一段时间，但并不属于文物范畴，言外之意就是：拆留均可。我们提出这个土地庙应该作为场地记忆得以保留，它是“物的原生”，是自然的一部分。土地庙是原住民生活的一种参与和创造，它在宗教心理、社会关系、村落起源、村落的景观格局等方面均体现出了与人的密切关系。因此，我们提出具体的修缮策略，并提出未来转化为社区图书馆、服务站继续承担连接邻里交往的场所功能。

Native Tree

We cannot forget the original power that has long gone away. Therefore, we proposed to the owner that in order to retain "homesickness", we should first preserve and transplant the existing trees to the maximum extent. They are the recorder and witness of the site, and they are also the hosts that accompany with the site. This strategy was quickly approved by the owners, but the subsequent problem was that due to the large underground garage planned for the building, the current trees would need to be transplanted nearby, and the cost of transplantation would be much higher than the cost of repurchasing seedlings. Moreover, since the construction season was in early summer, the survival rate of transplanting could not be guaranteed. So, move or buy? The owner is a little hesitant. Therefore, we proposed to compile different protection strategies according to the situation of different trees, number each tree, and propose transplantation and conservation methods one by one, so as to ensure the survival rate of transplanting and the lowest transplanting cost to the maximum extent. Our speciality make owners feel reassured, conservation and transplanting of existing trees can be realized. Finally, through continuous efforts, in the scope of first phase, 8 preserved trees are in situ (including two listed trees), more than 20 trees nearly transplanted. Over the years, every time we visited the project, and saw these protected trees, it's like old friends are reunited, and the rustle of leaves seems to thank us.

Native Content

There was a earth temple in the site. After the identification of the cultural relics department, it is believed that although the temple has been around for some time, it does not belong to the category of cultural relics. The implication is: demolition or retention both can be. We propose that this earth temple should be preserved as the site memory, it is "the native content", is a part of nature. Earth temple is a kind of participation and creation in the life of the indigenous people. It reflects the close relationship with people in religious psychology, social relations, village origin, and village landscape pattern. Therefore, we put forward specific renovation strategies, and proposed that in the future it will be transformed into a community library and a service station that will continue to serve as a place to connect neighbors.

五　自然做功

Natural Work

长辛店二期

Changxindian Phase ii

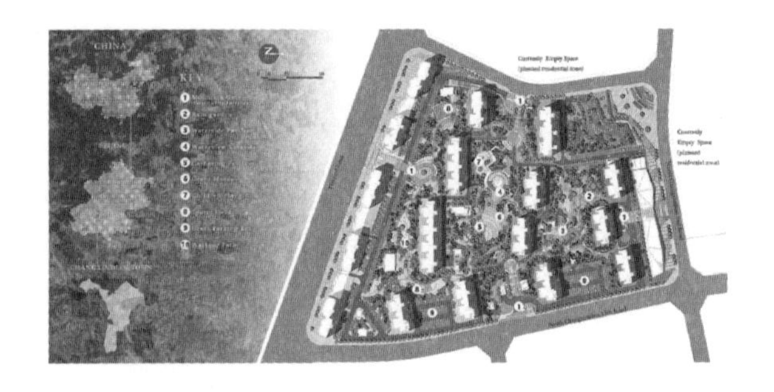

再续前缘

与中筑公司合作的十几年，我们完成了西府颐园、西局玉园、兆丰馨园、西府海棠等一系列项目，甲乙双方互相信任、互相尊重、互相成就，具有很好的合作基础。同时，最叫我们称道的还是十多年前该团队提出“打造中国最好的普通住宅”的企业目标，而且一步一步生根发芽、落地开花。

有了这些基础，在接到长辛店二期组团的设计任务时，我们和甲方便提出了更高的追求：如何以一种更加“高级”的方法落地“最好的普通住宅”，同时还能满足设计师尝试挑战不同手法的“私心”。

Once and Again

After more than ten years of cooperation with Zhongzhu, we have completed a series of projects such as Xifu Yiyuan, Xiju Yuyuan, Zhaofeng Xinyuan and Xifu Haitang. Both parties trust, respect and achieve each other, which has a good cooperation foundation. At the same time, we most commend the team that more than a decade ago, they put forward the corporate goal of "to build the best ordinary housing in China", and step by step take root, sprout and blossom.

With these foundation, when we received the group design task of Changxindian Phase ii, we and party A proposed a higher pursuit: how to complete the "best ordinary house" in a more "advanced" way, and at the same time to meet the "selfishness" of designers trying to challenge different approaches.

Wisely using rain gardens, bio-swale, bio-retention corridor, vegetation shallow ridges, and other infrastructures to prevent runoff from road surface pollution, slow down rainwater infiltration speed, and reduce flood risk caused by storms. After collecting and filtering by the LID system in the community, rainwater is reused in the wetland gardens, brooks, and artificial falls.

自然而然

我们决定在面对场地设计之前，先梳理一下合作的几个项目：从西府颐园的“古韵新作”到西局玉园的“文化挖掘”，再到兆丰馨园的“邻里景观”，再到西府海棠的“乡愁设计”，每个项目都提出了适宜的设计策略。这次如何用“新”来回应“最好的普通”呢？于是我们重新审视场地，翻阅10年前现场调研的照片：那自然原生的场地面貌唤起了我们对自然的美好回忆，巨大的自然感召力把我们的思维拉到了“自然”面前。对！就以“再造自然”为切入点，结合政府提出“海绵城市”建设的要求，我们决定利用整体场地南北巨大的高差设计贯穿全园的雨水收集系统，通过坡度计算设置汇水区域，再以模拟自然生境的方式布置雨水花园。这个设计方式的变化是对之前设计手法的一种“颠覆”，是从“心象”为主到“自然”为主的转变，空间形态不再受到主观“心象”的臆断，一切仿佛从土地上生长而出，一切自然而然。

Come Naturally

Before facing the site design, we decided to sort out several cooperative projects: from Xifu Yiyuan's "ancient with new advantages" to Xiju Yuyuan's "cultural excavation", then to Zhaofeng Xinyuan's "neighborhood landscape", and then to Xifu Haitang's "nostalgia design", each project has put forward appropriate design strategies. How does "new" respond to "best ordinary" this time? Therefore, we re-examined the site and looked through the photos of the site survey 10 years ago: the natural and original site appearance evoked our good memories of nature, and the great appeal of nature drew our thinking to "nature". Right! Taking "reengineering nature" as the starting point and combining with the government's requirements of "sponge city" construction, we decided to design the rainwater collection system throughout the whole garden by the huge height difference between the north and south of the whole site, set the catchment area through slope calculation, and then arrange the rain garden by simulating the natural habitat. The change of the design method is a kind of "subversion" to the previous design method, which is a shift from "mental image" to "natural". The spatial form is no longer subject to the subjective "mental image" assumption. Everything seems to grow out of the land, everything is natural!

然而，创作方法的转变是具有挑战性的，在北方的住区内设计如此之大的雨水花园，我们在设计之初对其最终效果及养护效果是底气不足的。巧的是，由于某些原因，项目竣工后，并没有能如期交付使用，而是推后验收。而这一年的时间居然成了我们景观设计的检验期。实践证明，在几乎没有养护的情况下，园内植物生境完美呈现，景观效果及生态效益得到彰显，甚至说是超过预期效果，我们可以非常肯定地说做到了“设计结合自然，设计让自然做功！”

However, the change in creative methods was challenging, and our confidence that design such a large rainwater garden in the residential areas of the northern region was not enough to ensure its final effect and conservation effect at the beginning of the design. Coincidentally, for some reason, the project was not delivered as scheduled after completion, but delayed acceptance. And this one year time unexpectedly became our landscape design inspection period. Practice has proved that, in the case of almost no conservation, the plant habitat in the garden is perfect, the landscape effect and ecological benefits are highlighted, even more than the expected effect, we can be very sure to say that "design combined with nature, design let nature do the work!"

六　植物营境

Plants Construct Environment

青岛德国中心

German Center in Qingdao

植物美境

植物在园林景观中既有静态之美也有动态之美，既有一时之美，也有四季之美。在青岛德国中心的景观设计中，我们就用植物营造了四季的美景。这里的春天，湖风驯草；这里的夏天，繁花摇曳，蜂飞蝶舞；这里的秋天，橙黄褐木，果色斑驳；这里的冬天，苍枝遒劲，松映静湖。

Plants Beautify Habitat

Plants in the garden landscape has both static beauty and dynamic beauty, both temporary beauty, but also the beauty of the four seasons. In the landscape design of German center in Qingdao, we created the beautiful scenery of the four seasons with plants construction. Here in spring, the wind from lake sways grass; here in summer, flowers in bloom sway, and bees fly together with butterflies dance; here in autumn, trees are in orange and yellow, fruit color are mottled; here in winter, green branches are strong, pine reflects in the quiet lake.

群落生境

青岛德国中心坐落于原有生境条件良好的水库之畔，人为的建设与生态恢复要同时兼顾，作为原有生态系统的补偿，植物在这片土地上是生命生生不息的源泉。植物群落的差异在园子里体现为陆地群落、半湿生群落和湿生群落。陆地群落的设计充分考虑了乔灌木的差异化、色彩季相的差异化，通过人工与半人工的组合形成层次丰富的植被格局。场地到水库边的植被缓冲带，充分发挥了地被层的净化作用，地被植物的搭配以植物生长习性相近为原则，根据生境与水库的远近关系，将地被划分为四种类型，对应的生境分别为干-半干、半干-半湿、半湿-潮湿、湿生。整个植被缓冲区以生境特征为设计依据，综合考虑景观使用功能和视线组织要求，形成了富于静态美和动态美的自然的群落生境。

Community Build Habitat

Germany center in Qingdao is located beside the reservoir that has original habitat in good conditions, and artificial construction and ecological restoration should be considered at the same time. As a compensation for the original ecosystem, plants in this place are the source of life. The difference of plant community is reflected in the land community, semi-wet community and wet community in the garden. In the design of the land community, the differentiation of arbor and shrub and the differentiation of seasonal phase of color were taken into full consideration, and the combination of artificial and semi-artificial forms the rich vegetation pattern. The vegetation buffer zone from the site to the edge of the reservoir gives full play to the purifying effect of the ground cover. The matching of ground cover plants is based on the principle of similar plant growth habits. According to the relationship between the habitat and the reservoir, the ground cover is divided into four types. The corresponding habitats were dry - semi - dry, semi - dry - semi - wet, semi - wet - wet and wet. The whole vegetation buffer zone is designed on the basis of habitat characteristics, considering the landscape function and the requirements of line of sight, forming a natural community habitat rich in static and dynamic beauty.

南宁园博园东南亚特色植物园

Southeast Asia Characteristic Botanical Garden in Nanning Garden Expo

异域的“特色”

经验告诉我们，很多设计在进展过程中都会发生变化。南宁园博园东盟园也是如此，在项目已经开工建设后，我们突然接到了业主的“新任务”：整合原东盟馆周边景观，结合东盟园展区打造东南亚特色植物园。

现场已经开工，距离开园还有一年不到的时间，大量的现场配合与调整已经把我们压得喘不过气来，这无疑对于我们来说是雪上加霜。经过与业主的沟通，我们也理解了这个改变的目的。简言之，就是要做“特色中的特色”，要全面展示东南亚景观特色，打造国内最具特点的东南亚特色植物园，作为面向东盟十国展现友好共生理念的窗口。

此时周边场地已经实施了很多，有崔愷院士领衔设计的东盟馆，有东盟十国设计师不同风格、代表各自国家特色的东南亚展园，还有复杂多变的地形。那么，如何在如此“异域”的环境中做出特色？如何统领全区？如何反映主题？植物！植物！只有植物！既能让一切不同风格、不同材质、不同尺度的建构筑物相互融合，又能扣“特色植物园”的主题，还能呼应东盟诸国和谐共生的理念。

Exotic "Features"

Experience tells us that many designs change as they progress. The same is true for the ASEAN Garden of Nanning Garden Expo. After the project has been started, we suddenly received a "new task" from the owner: integrating the surrounding landscape of the original ASEAN Garden and combining the ASEAN Garden exhibition area to create a botanical garden with southeast Asian characteristics.

The site has been started, and there is less than a year before the opening of the park. A large number of on-site cooperation and adjustment has put us under huge pressure, and it is undoubtedly for us to add insult to injury. After communicating with the owner, we also understood the purpose of this change. In short, it is to do the "characteristics in characteristics", to fully display the landscape characteristics of southeast Asia, to create the most featured southeast Asia characteristic botanical garden in China, to be as a window for the ten ASEAN countries to show the concept of friendly symbiosis.

At this time, a lot of surrounding sites have been implemented, including the ASEAN pavilion design led by academician Cui, the southeast Asia exhibition garden with different styles of designers from the ten ASEAN countries representing the characteristics of their respective countries, and the complex and changeable terrain. So, how to make a feature in such an "exotic" environment? How to govern the district? How to reflect the theme? Plant! Plant! Only plants! It can not only make all the structures with different styles, materials and scales blend with each other, but also adopt the theme of "characteristic botanical garden" and echo the concept of harmonious coexistence among the ASEAN countries.

植物的“特色”

在特色营造上，为了与十国“手拉手”的东盟馆建筑造型相呼应，在东南亚展园区利用地形，形成心形环路串联十园，并以抽象提取的花形平台俯瞰东南亚展园，寓意“心连心”。同时围绕各国国花主题，抽象提取国花形象，提出“携手连心，国花共生”的主题概念。

"Characteristics" of Plants

In terms of characteristic construction, in order to echo the architectural shape of the ASEAN pavilion with ten countries "holding hands", the terrain of the southeast Asia exhibition park is utilized to form a heart-shaped ring connecting the ten gardens, and the flower shaped platform is abstracted to overlook the southeast Asia exhibition garden, implying "heart to heart". At the same time, focusing on the theme of national flower of each country, the image of national flower is abstractly extracted, and the theme concept of "hand in hand with heart, national flower symbiosis" is put forward.

在种植设计上，我们与南宁古今园林院通力合作。南宁地处桂南地区，属南亚热带季风性气候，天然植被类型为“季节性雨林”，东南亚地区的植物品种大部分在该地能良好地生长，部分特色植物建议引种使用。最终确定“茂林繁花，层林尽翠”的植物种植主题，通过展示东南亚地区各国的特色植物，营造独具特色风貌的东南亚热带植物景观。

整个植物园分为东南亚特色种植区、水畔棕林种植区、纪念林种植区、特色林种植区：

东南亚特色种植区：展示各国国花或名花、国树或名树，东南亚国家花卉品种丰富，其代表性的花、树并不单一。包括中国的牡丹、文莱的“辛波嘎加”（康定杜鹃）、柬埔寨的隆都花、印度尼西亚的毛茉莉、老挝的占巴花（鸡蛋花）、马来西亚的扶桑花、缅甸的龙船花、菲律宾的毛茉莉、新加坡的‘卓锦’万代兰、泰国的金链花、越南的莲花。

水畔棕林种植区：结合现有的滨湖景观和东南亚建筑元素，植物选择凸显亚热带氛围，并作为水体、建筑元素的融合过渡，打造“水畔棕林”的植物景观特色。亲水区域以“睡莲”作为植物主题，种植“红色系”睡莲，营造热烈大气的睡莲展示区域。上层以斜飘特型棕榈科植物为主，营造自然野趣的湿地空间。配景植物：特型椰子、特型银海枣、水生美人蕉、沼地棕、旱伞草、蝎尾蕉、鱼尾葵、董棕、三药槟榔 、香桄榔、矮蒲葵等。

In planting design, we cooperate with Nanning Gujin Institute of Landscape Planning & Design. Nanning is located in the Guinan region, belongs to the south of subtropical monsoon climate, and the natural vegetation type is "seasonal rain forest". Most of the plant species in southeast Asia can grow well in the local area, and some characteristic plants are recommended to be introduced and used. Finally, the plant planting theme of "luxuriant forests with flowers and lush forests with trees" was determined. By displaying the characteristic plants of southeast Asian countries, the landscape of tropical plants with unique features in southeast Asia was created.

The whole botanical garden is divided into the planting area with southeast Asian characteristics, the planting area of water-side palm forest, the planting area of memorial forest and the planting area of characteristic forest:

Planting area with southeast Asian characteristics: Display national flowers or famous flowers, national trees or famous trees from every country. Southeast Asian countries are rich in flower varieties, and its representative flowers, trees are not single, including China's peony, Brunei "Xinbo Gajia" (kangding cuckoo), Cambodia's rumdul, Indonesia's jasminum multiflorum, Laos' Champa flower (frangipani),Malaysia's hibiscus rosa-sinensis, Burma's chinese ixora flower, Philippines' jasminum multiflorum, Singapore's "ZhuoJin" Vanda, Thailand's laburnum, Vietnam's lotus.

Planting area of water-side palm forest: combined with the existing lakeside landscape and southeast Asian architectural elements, the selection of plants highlights the subtropical atmosphere, and as a transition of water and architectural elements, create the plant landscape characteristics of "water-side brown forest". The water-loving area takes "water lily" as the theme of the plant, planting "red" water lily to create an enthusiastic display area of water lily. The upper layer is dominated by special oblique palm plants, creating a natural and wild wetland space. Collocation plants with landscape: special coconut, special type silver date, canna generalis, acoelorrhaphe wrightii, cyperus alternifolius, heliconia, fishtail palm, Dong palm;areca triandra, Arenga tremula, Livistona humilis and so on.

纪念林种植区：结合优越的湖湾景观以及东盟馆、金色大厅等建筑元素，把植物作为热带建筑不可缺少的立面元素，配合建筑营造“纪念林”植物景观，种植代表东南亚十国友谊的友谊树，如十国国树、名树，形成东南亚植物园独具特色的纪念林。如中国南宁的扁桃（市树）、越南的木棉（国树）、老挝的鸡蛋花（国树）、泰国的桂树（国树）、缅甸的柚木（国树）、印度尼西亚的檀木、文莱的菩提树、马来西亚的橡胶树、柬埔寨的糖棕、新加坡的紫檀、菲律宾的纳拉树。

特色林种植区：作为全园的种植背景，在基础绿化种植的基础上增加种植层次、林缘花带，以东南亚地区特有的佛教礼仪植物“五树六花”作为种植主题，烘托园区整体的东南亚景观氛围。东南亚的园林在很大程度上都受到佛教的影响，植物常常与宗教联系紧密，具有强烈的象征意义。在东南亚的佛教园林中，“五树六花”是必不可少的。

Planting area of memorial forest: combining the superior bay landscape with the architectural elements of the ASEAN pavilion and Golden Hall, plants are regarded as the indispensable facade elements of tropical architecture, and the "memorial forest" plant landscape is built in cooperation with the architecture. Friendship trees representing the friendship of the ten southeast Asian countries, such as the national tree of the ten countries and famous trees, are planted to form the unique memorial forest of the southeast Asian botanical garden, such as almond from Nanning, China (city tree), kapok from Vietnam(national tree), Frangipani from Laos (national tree), laurel from Thailand (national tree), Burmese teak (national tree), sandalwood from Indonesia, banyan tree from Brunei, rubber trees from Malaysia, borassus flabellifer L.from Cambodia, red sandalwood from Singapore, Narayanan tree from Philippines.

Planting area of characteristic forest: as the planting background of the whole garden, the planting level and the flower belt along the forest edge should be increased on the basis of greening and planting. The Buddhist ritual plant "five trees and six flowers" which is unique to southeast Asia is used as the planting theme, highlights the overall southeast Asian landscape atmosphere of the park. Gardens in southeast Asia are heavily influenced by Buddhism, and plants are often closely associated with religion and have strong symbolic meanings. In the Buddhist gardens of southeast Asia, "five trees and six flowers" are indispensable.

第四章

文明的表现

Chapter Four

Manifestation of Civilization

一 场所再生
Place Regeneration

拉萨布达拉宫周边及宗角禄康公园

Surroundings the Potala Palace in Lhasa and Zongjiao Lukang Park

复杂的现状

在布达拉宫北坡山下有一座公园，藏语称“禄康插木”，是拉萨市民休闲娱乐、布达拉宫旅游者疏散接待、联系拉鲁湿地和拉萨河生态链的重要场所。然而，公园现状设施陈旧、景点缺乏、游人稀少，周边被沿街的办公和商业建筑遮挡，使得行人无法领略布达拉宫和公园的景观，导致公园虽身处要地却几乎被世人遗忘。

文明的策略

景观改造不能就场地论场地，尤其对于敏感的布达拉宫周边，需要从历史上的布达拉宫周边环境、现在的布达拉宫周边环境、城市空间性质、城市发展形态、城市发展性质、视廊系统、用地功能、文物保护要求等方面进行系统分析，回答为什么要整治与改造。通过对布达拉宫周边错综复杂的环境的理性分析后，我们认为整治与改造工作不可能一步到位，它必须要经历一个分步实施、逐步改善和恢复的过程。因此，“有机生长”是适合场地特征的文明策略，公园“二期四步”的分步实施，将会使这个场地有机生长成拉萨市民的城市客厅。

Complex Situation

At the foot of the north slope of the Potala Palace, there is a park, which is called "Lu Kang Cha Mu" in Tibetan. It is an important place for the leisure and entertainment of Lhasa residents, the evacuation and reception of Potala Palace tourists, and the connection of ecological chain between the Lalu Wetland and the Lhasa River. However, the existing facilities of the park are old, scenic spots are scarce, and there are few visitors. The surrounding area is blocked by the office and commercial buildings along the street, which makes the Potala Palace and the park inaccessible to pedestrians. As a result, although the park is in an important place, it is almost forgotten by the world.

Civilized Strategy

Landscape transformation can't just consider the site as it stands, especially for sensitive surroundings around the Potala Palace.The surrounding environment of Potala Palace in history, the surrounding environment of Potala Palace now, the urban spatial character, urban development form, urban development character, corridor system, land use function, cultural relic protection requirements and other aspects need to be systematically analyzed, and answer the question why we need renovation and transformation. Through the rational analysis of the complex environment around Potala Palace, we believe that the renovation and transformation work can not be completed in one step, and it must go through a step-by-step implementation, gradual improvement and recovery process. Therefore, "organic growth" is a civilized strategy suitable for the characteristics of the site. The implementation of "two phase and four steps" of the park will make the site grow organically as the urban living room of Lhasa residents.

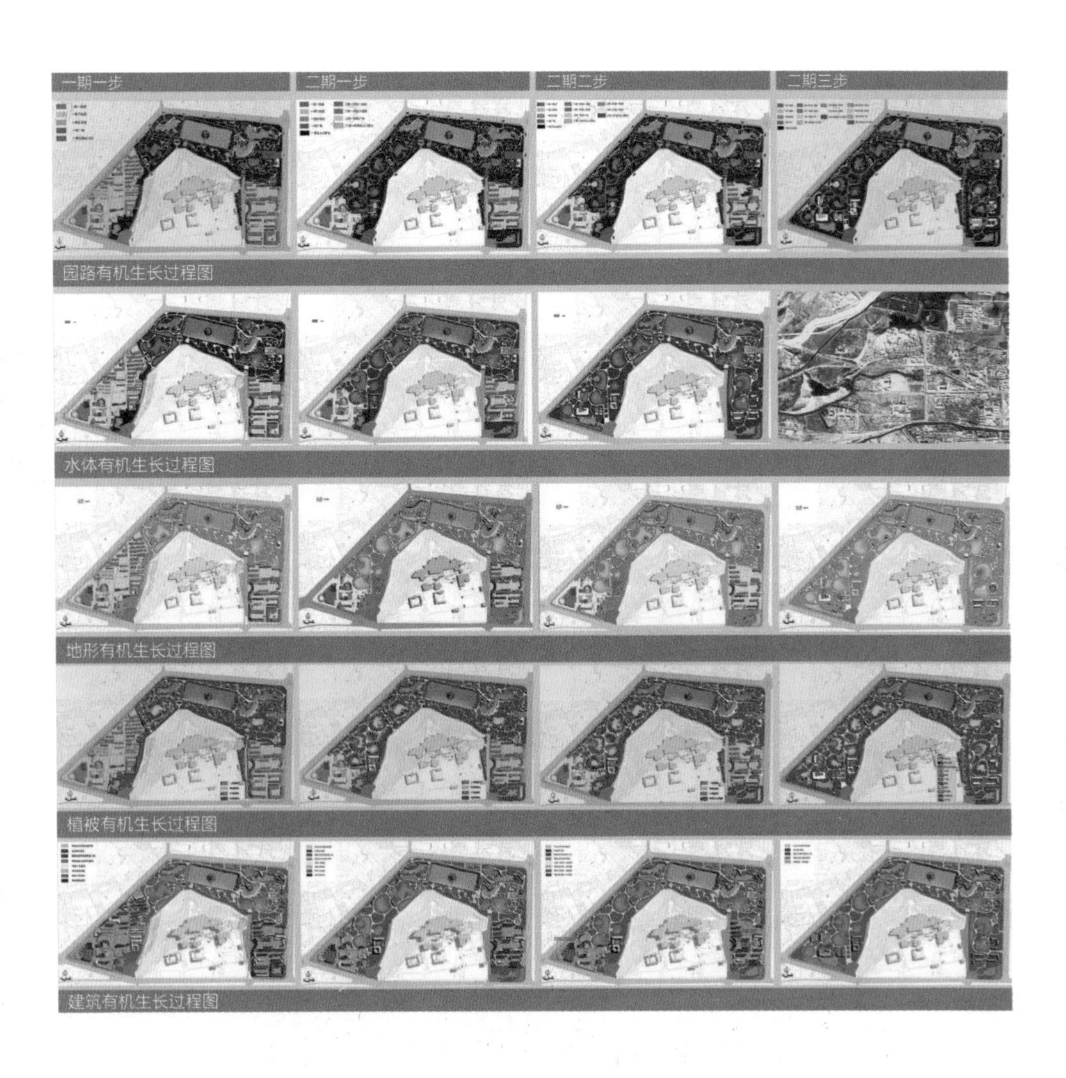
一期一步
二期一步
二期二步
二期三步
园路有机生长过程图
水体有机生长过程图
地形有机生长过程图
植被有机生长过程图
建筑有机生长过程图

场所的再生

我们有计划、有步骤地疏理、整治、更新布达拉宫周边的环境，使其真实反映历史风貌，并留出发展的可能性，始终保持布达拉宫周边环境的可持续发展态势，让场所实现有机再生，让文明得以延续。公园改造以突出布达拉宫以及场地内龙王庙、白塔等遗迹遗存为主，摆正了与布达拉宫的空间关系，提升了世界文化遗产周边环境的品质，保留现状成片的树林和水边特色的左旋柳，尊重历史环境的同时，强化了公园的生态功能。在保留原有龙王潭的基础上，通过水系的营造将公园原有分散的水面串联成为一个整体，打造了蓝绿交织的亲水空间。

Regeneration of Site

We sort out, renovate and renew the surrounding environment of the Potala Palace as planning and step by step, so that it can truly reflect the historical landscape, and leave the possibility of development. The sustainable development of the surrounding environment of Potala Palace has always been maintained, so that the site can realize organic regeneration and the continuation of civilization. The renovation of the park is mainly to highlight the Potala Palace and the remains of the dragon king temple and white pagoda in the site, straighten out the space relationship with Potala Palace, improve the quality of the surrounding environment of the world cultural heritage, and retain the existing forest and the characteristic paraplesia var. Subintegra near waterside. While respecting the historical environment, it strengthens the ecological function of the park.On the basis of retaining the original Longwangtan, the original scattered water surface of the park is connected as a whole through the construction of water system, creating a blue-green interwoven hydrophilic space.

二　文明再现

Civilization Reappearance

西安未央宫国家遗址公园

National Heritage Park in Xi 'an Weiyang Palace

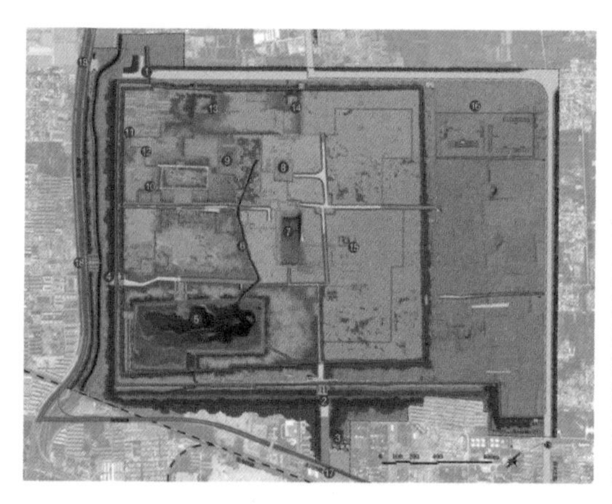

体现场地在自然演替中的初始价值

汉长安城未央宫遗址是丝绸之路的起点，将其作为遗址公园展示是丝绸之路成为世界文化遗产的一个很重要的支撑。在设计之初，我们就有一个非常明确的原则：即梳理能够体现未央宫价值特征的空间表达，这种“明确”使我们无法采用非遗址公园类项目设计中“开放式和随意性的操作”模式。严谨考据成为设计必须保持的态度。记录现状每一类植被品种、生长状态及分布情况，梳理遗址现状及整体空间格局，研究历史时期的价值信息，根据历史信息获得有效数据等成为设计的重点。在此基础上，我们首先通过保留、调整、迁移、清除四项措施对现状植被进行梳理；其次，采用植物的勾边、打底、选配等设计手法构建遗址整体格局与氛围；再次，利用简洁、自然的设计手法对游客服务场地及设施进行设计，在材料的选配上选择符合遗址文化、色彩特征的材料。

Reflect the Initial Value of the Site in Natural Succession

The site of Weiyang Palace in Chang 'an City in Han Dynasty is the starting point of the silk road. Displaying it as a heritage park is an important support for the silk road to become a world heritage site. At the beginning of the design, we have a very clear principle: that is, combing the space expression that can reflect the value characteristics of Weiyang Palace. This "clarity" precludes us from adopting the "open and arbitrary operation" model of non-heritage park design. Rigorous examination has become an attitude that must be maintained in design. It has become the focus of the design to record the species, growth status and distribution of each type of vegetation, sort out the current situation and overall spatial pattern of the site, study the value information of the historical period, and obtain effective data based on the historical information. On this basis, we firstly sorted out the current vegetation through four measures: reservation, adjustment, migration and clearance. Secondly, the whole pattern and atmosphere of the site were constructed by the design methods of hooking the edge, making the base, optional match and so on. Thirdly, the tourist service sites and facilities are designed with simple and natural design methods, and materials matching with the cultural and color characteristics of the site are selected.

漢長安城未央宮遺址

约定现场工作的五重措施

在实施过程中，我们用五重措施约定了设计师与施工方的现场工作，用以保证遗址空间格局的完整性与本真性，恢复特定历史环境并融入展示功能。一是设计师在现场与施工方明确现状保留植被品种并对其进行坐标定位；二是设计师在现场要求施工方除常规种植要求外，遗址公园内植物的种植以不对遗存造成任何破坏为前提和基本原则，充分考虑植被对遗址本体的影响并进行评估，并严格按照苗木表中的苗木品种、规格、种植方式实施；三是设计师与施工方明确禁止动土、限制动土、控制动土及无扰土限定范围及种植要求；四是设计师与施工方一起赴苗圃或山野区寻找符合遗址环境的植被，并在可能的情况下在现场周边进行假植，以利于植被的生长与存活；五是设计师现场确定避让现状场地中地下管线及地面构筑物的原则，并明确最小水平距离。

正是由于这种严谨的“新”设计理念，跳出了一般设计师墨守成规的职业设计思想，脱离了自由畅想主义的“困境”，实现了遗址公园特有的“创新”。

Appoint Five Measures for Site Work

In the process of implementation, we agreed on the site work of the designer and the construction party with five measures to ensure the integrity and authenticity of the spatial pattern of the site, restore the specific historical environment and integrate the display function. First, the designer made clear with the construction party about the current situation of preserving vegetation varieties in the site and made coordinate positioning for them; second, in addition to conventional planting requirements, the designer requires the construction party to plant plants in the site park on the premise and basic principle that no damage will be caused to the remains, fully consider the impact of vegetation on the site itself and evaluate it, and strictly follow the seedling varieties, specifications and planting methods listed in the seedling table; third, the designer and the construction party clear and definite the limit range and planting requirements about forbidding moving the soil, limiting moving the soil, controlling moving the soil and not disturbing the soil; fourth, the designer and the construction team go to the nursery or mountain area to find the vegetation suitable for the site environment, and plant around the site if possible, so as to facilitate the growth and survival of the vegetation; fifthly, the designer shall determine the principle of avoiding the underground pipelines and ground structures in the current site and specify the minimum horizontal distance.

Because of this kind of rigorous "new" design concept, we jump out of the hidebound career design thought of general designers, break away from the "dilemma" of free imagination, realize the unique "innovation" of the relic park.

三　守望传承

Watch for the Inheritance

湖南老司城遗址公园

Laosicheng Ruins Park in Hunan

守护与传承是必然遵循　每一处遗址的发现都对研究特定历史阶段的人类文明具有重要价值，面对遗址公园的设计，真实性的保护展示与文化的守护传承就成为我们必须要遵照的原则。

Guard and inheritance is inevitable to follow

The discovery of each site is of great value to the study of human civilization in a specific historical stage. Faced with the design of the site park, the protection and display of authenticity and the protection and inheritance of culture have become the principles we must follow.

展示与利用是新的赋予

我们所做的努力是从遗址的保护、展示及利用三个方面探究遗址环境设计的特点及方法，使历史环境、遗址格局与现有的村落风貌、自然环境之间能获得一个平衡点；同时，我们总结遗址环境设计中本体、环境、生态、景观、人等方面的关系，探究遗址文化与现代文明邂逅的保护、展示与利用问题，构建一个适应的、平和的、安静的遗址界面，使其既守护遗址又融入现有的生活状态，使环境获得新的生命力。

植根乡土回归自然

我们的设计以山水格局为依托，植根乡土、回归自然。遗址博物馆周边自然条件与生态基底是景观的核心，为避免对区域山水格局造成破坏，景观采取最小干预的造园手法；以自然山水为展示对象，减少人工雕琢和大规模的土方开挖，将景观、建筑与环境相融合；随着地势高低变化，利用卵石墙、植物及小桥进行空间的连接，同时利用地势及水位高差，理石散流，构成小型叠水，使景观野趣横生。

Display and Utilization Are New Gifts

What we try to do is to explore the characteristics and methods of the environmental design of the site from three aspects: the protection, display and utilization of the site, so as to strike a balance between the historical environment, the site pattern, the existing village style and the natural environment. At the same time, we summarize the relationship among ontology, environment, ecology, landscape and human in the environmental design of the site, explore the protection, display and utilization of the site culture and modern civilization encounter, and build an adaptive, peaceful and quiet site interface, so that it not only protects the site but also integrates into the existing living state, so that the environment can gain new vitality.

Take Root in Rural Areas and Return to Nature

Our design is based on the pattern of mountains and river, take root in rural areas and return to nature. The surrounding natural conditions and ecological base of the site museum are the core of the landscape. In order to avoid the damage to the regional pattern of mountains and river, the landscape adopts the gardening method of minimal intervention. With natural mountains and river as the display object, reduce artificial carving and large-scale earthwork excavation, and integrate landscape, architecture and environment; with the change of terrain, pebble walls, plants and small bridges are used to connect the space, while the terrain and water level difference are also used to make orderly stones and scattered flow, forming small overlapping water, wild and interesting.

心象自然篇

Inner Nature Imagery

心象自然

对设计师而言，“心象”和“自然”是影响设计的两个方面，一个是主观的、内在的，一个是客观的、外在的。两个方面在不同的项目中往往侧重不同，在功能性要求高或艺术性表达强烈的项目中，往往偏重“心象”的表达。而在项目场地中自然因素比较突出或生态要求比较高的时候，“自然”的呈现就变得更加重要。

当然，这也不是绝对的，在相对综合的项目中，设计作品有时很难分清是“心象”呈现得多还是顺应“自然”的多，也许某个区域以自然为主，某个区域以人工为主。也许自然中有人工，人工中有自然，你中有我，我中有你，互相融合。《园冶》中“虽由人作、宛自天开”，大概就是这样的境界。

从创作观念上看，“心象自然”中的“象”还可以做动词，既内心以自然为“象”，创作中将自然内化于心，再由心外化于景观。对自然进行二次创作，所谓“象为自然，与心相应”，而这也许就是我们要追求的一种状态。

Inner Nature Imagery

As designers, “inner expression” and “nature” are the two aspects that influence our design the most. One is subjective and internal, the other is objective and external. These two parts tend to focus on separate aspects in different projects. Designers focus more on expressing inner thoughts in the case requiring higher function or artistic creation; and in the project including more natural elements, it is more important for our designers to represent nature.

Certainly, nothing is absolute. In some relatively more comprehensive projects, it is hard to tell if its more inner expression or nature following. Maybe some areas are more natural-based, but others have more artificial designs. Man live in nature, and nature evolves with man. Man and nature is a mutual integration. Just like what “Yuanzhi”, an ancient Chinese gardening book, describes “Work by man, like god did”.

In creation aspect, the Expressed in “Inner Nature Imagery” can be a verb. When working on a project, designers switch nature into an inner feeling, and then externalize it back on landscape design. We make second creation of nature, like what we are pursuing of— “Expressed nature, echo our heart”.

第五章

心象与自然的平衡

Chapter Five

Balance Between Mind and Nature

一 建筑景观一体建构

Integrated Construction of Architecture and Landscape

1. 景观建筑

Landscape Architecture

浙江大学生命学院

College of Life in Zhejiang University

门里门外与人工自然

有时候在想，建筑和景观的区别在什么地方？如果以门为界，门内是建筑，门外是景观，建筑偏人工，景观偏自然。建筑是人造物，虽然在自然中，但要更多考虑人的需要，可以回避自然，可以唯建筑，要突出以人为本。而景观是人造自然，以自然为对象，以自然为手段，回避不了自然，要么与自然形成对比，要么顺应自然，天人合一。

你中有我，我中有你

建筑与景观又是密不可分的，大建筑的观念认为，景观是建筑的一部分，从大景观的视角看，建筑是景观的重要组成。其实本没有什么绝对的界限，如果有，那也是你中有我，我中有你。所以设计师将建筑和景观一体化，使二者互为融合，相得益彰，就显得非常重要。

Inside and Outside the Door and Artificial Nature

Sometimes we wonder, what is the difference between architecture and landscape? If the door is the boundary, the inside of the door is the architecture, outside is the landscape, the architecture is artificial, the landscape is natural. Architecture is artificial object. Although it is in nature, the needs of man should be more considered, can avoid nature, can only think of architecture, to highlight the people-oriented. Landscape is artificial nature, with nature as the object, with nature as the means, can not avoid nature, either form a contrast with nature, or conform to nature, be the unity of heaven and man.

There's Me in You, There's You in Me

Architecture and landscape are inseparable. The concept of big architecture believes that landscape is a part of architecture, and from the perspective of big landscape, architecture is an important part of landscape. Actually it does not have absolute limit, if have, that is, there's me in you, there's you in me. Therefore, it is very important for designers to integrate architecture and landscape so that the two are mutually integrated and complement each other.

突出景观的建筑

2002年，崔总接到浙江大学生命科学学院项目的设计委托，因为涉及生物实验室等有洁净要求的功能，于是找到正在做四川省人民医院项目的我，由此我与崔总的合作便结下了缘分，由建筑合作到后来很多建筑与景观的合作。而浙江大学生命科学学院的建筑设计，在解决了基本建筑功能问题后，更多的是以突出景观为主在展开设计。与环境协调并关注景观营造，使得生命科学学院成为一个景观建筑。

Building That Highlights the Landscape

In 2002, academician Cui was commissioned to design by the college of life sciences of Zhejiang University. Because it involved the biological laboratory and other functions required for cleanliness, I was found when I was doing the project of Sichuan Provincial People's Hospital. Therefore, the cooperation between me and academician Cui resulted in a fate, from architectural cooperation to many architectural and landscape cooperation later. The architectural design of the college of life sciences of Zhejiang University, after solving the problem of basic architectural functions, focuses more on highlighting the landscape design. Coordinating with the environment and focusing on landscape construction, the college of life sciences becomes a landscape architecture.

从环境中长出的建筑

新的生命科学学院位于浙江大学紫金港校区的西南角，西侧和南侧有护校河从外围流过，环境清幽。场地东西长南北短，于是布局以一幢东西向的主体建筑为主干，向南向北伸出枝状的附属建筑，体现一种能生长的、可持续发展的意向，也符合生命科学学院的属性。

建筑本身就是校园西南角的景观，整个建筑主从分明，体量错落有致、变化丰富。有1层的报告厅，3层通高的门厅和玻顶大厅以及4层、5层、局部6层的办公用房和各类实验室。建筑形态力求现代、简洁、富于韵律、逻辑关系清晰。在材料的运用上主体建筑以灰砖为主，而枝干建筑则以白色墙板与主体形成对比，在楼电梯及局部凸出体量的处理上则运用了钢框架及玻璃。电梯的设计成为整个建筑的标志，透过黑色钢框架及半透明的玻璃外表可以看到红色的电梯构件及轿厢，随着红色轿厢的上下移动，一种生命的律动感得以展现，从而与生命科学学院的内质取得呼应。

Buildings That Grow out of the Environment

The new college of life sciences is located in the southwest corner of Zijingang campus of Zhejiang University, the west side and the south side have the school river flows from the periphery, and the environment is quiet. The site is long from east to west and short from north to south, so the layout is mainly composed of an east-west main building and a twin-shaped subsidiary building extending from south to north, reflecting the intention of growth and sustainable development, which also conforms to the attributes of the college of life sciences.

The building itself is the landscape of the southwest corner of the campus. The whole building make a distinguish differentiate between the principal and subordinate, with well-arranged volumes and rich changes. There is a lecture hall on the first floor, a hall with full-height foyer and a hall with glass roof on the third floor, as well as offices and laboratories on the fourth, fifth and partial sixth floors. The architectural form strives to be modern, concise, full of rhythm and clear logical relationship. In terms of the use of materials, the main building is mainly made of grey bricks, while the branches building is made of white wall panels to form a contrast with the main body. In terms of the treatment of building elevators and partial protruding volumes, steel frames and glass are used. The design of the elevator becomes the symbol of the whole building. Through the black steel frame and translucent glass appearance, the elevator components and cage are red. With the movement of the red cage up and down, a sense of rhythm of life is displayed, thus echoing the essence of the college of life sciences.

由建筑延伸出的景观

建筑内外的景观营造也是设计的重点。在主入口大厅的对面，有一个玻璃顶大厅，供教师和学生休息使用，大厅内除了提供茶点和咖啡外，还营造出植物温室的景观场景，阳光透过玻璃，婆娑的树影与休憩的人影交相辉映，人与自然共处一室，体会看与被看的乐趣。

玻璃顶大厅的景观一直延伸至南侧的室外，这里设计有一处静水池，将护校河的水引入其中，建筑的倒影与水中的雕塑平台使得南侧的小广场富有吸引力。

在南侧水池的两边是学院的后花园，结合学院生命科学的实验需要，这里被规划为一块块植物试验田，呈现出模数化的肌理，室外温室与小块休息场地穿插其间，满足实验与游憩的双重功能，独特的景观也标识出建筑的特殊气质。

生命与科学

2003年的春天，正是非典盛行的时候，也是生命科学学院设计出图的时候。带着对生命的敬畏和对建筑的执着，我们没有休息，印象最深的场景是戴着口罩和崔总对图，带着一点点对非典的恐惧在深夜里画图。这也许是兼顾生命与科学的一次设计吧。

Landscape That is Extended from the Building

Landscape construction inside and outside the building is also the focus of the design. Opposite the main entrance hall, there is a glass-topped hall for teachers and students to rest. Besides providing refreshments and coffee, the hall also creates a landscape scene of a greenhouse.The sunlight passes through the glass, the swaying shadows of the trees and the figures at rest mingle with each other, the person and nature coexist a room, experience the pleasure of seeing and being seen.

The view of the glass-topped hall extends to the outdoor space of the south side, where a standing pool is designed to draw the school river water from outside, and the reflection of the building and the sculpture platform in the water make the small square on the south side attractive.

On either side of the pool on the south side is the back garden of the college. In combination with the experimental needs of life science of the college, it is planned to be plant experimental fields with a modular texture. Outdoor greenhouses and small rest areas are interspersing with each other to meet the dual functions of experiment and recreation. The unique landscape also marks the special temperament of the building.

Life and Science

In the spring of 2003, when SARS was in full swing, the college of life sciences was coming up with the design drawing. With a reverence for life and an insistence to architecture, we did not rest. The most impressive scene is checking the drawings with academician Cui and wearing a mouth-muffle, drawing sketches in the middle of the night with a little fear of SARS. Perhaps this is a design both for life and science.

永城日月湖景区游客服务中心

Tourist Service Center of Riyue Lake Scenic Spot in Yongcheng

景区内的建筑

永城是河南商丘的一个县级市，因汉高祖刘邦在域内芒砀山斩蛇起义而被誉为“汉兴之地”，这里也是全国六大无烟煤基地之一，多年的煤炭开采形成了很多采空区，地面不同程度下陷，地下水露出，形成类似沼泽的地表形态，而日月湖景区正是这样一个生态修复性的项目，经过八年多由规划到实施的过程，挖湖推山、筑路搭桥、因势造景，如今24平方公里的日月湖景区已初具规模。景区内的游客服务中心，是一个值得述说的景观建筑，与大多数建筑为主、景观为辅的思路不同，本建筑是从景观中长出来的，是景观中的建筑。那么，景观中的建筑有哪些特质呢?

Buildings in the Scenic Area

Yongcheng is a county-level city in Shangqiu, Henan Province, known as the "land of Hanxing" because of the Han Gao-zu Liubang's uprising by killing snakes on Mount Mangdang in the region. It is also one of the six largest anthracite coal bases in China. Many gobs have been formed by years of coal mining, the ground sinks to different degrees, and underground water emerges to form a surface configuration like marshy. The Riyue Lake scenic spot is just such an ecological restoration project. After more than eight years,from planning to implement the process, digging lakes and pushing mountains, building roads and bridges, creating landscape by situation, now 24 square kilometers of the lake scenic spot has begun to take shape. The tourist service center in the scenic spot is a landscape architecture worth telling about. Different from the many ideas that buildings are primary and are supplemented by the landscape, it grows out of the landscape and is the architecture in the landscape. So, what are the characteristics of architecture in the landscape?

建筑是用来看的

新的建筑方针是“适用、经济、绿色、美观”，而对这个建筑来说，形态和气质的塑造是第一位的，也就是说，美观似乎更重要，它更像是用来看的，它一定要与场地的特征相契合才对。建筑场地是在景区的艺术片区，位于椭圆形的大草坪的一端。大草坪像蛋糕一样被网格状的小路分割成很多不规则的三角形或梯形，在建筑周围，这些小路有所下沉，于是建筑和花池便凸现了出来，成为一组高高低低的被切割的形态，这组形态成为大草坪的中心，成为艺术区中的艺术品。

Architecture Is For Viewing

The new architectural policy is "applicable, economical, green and beautiful", and for this building, the mould of shape and temperament are the first. That is to say, beauty seems to be more important, it is more like to be used for seeing, and it must fit with the characteristics of the site. The construction site is located in the art section of the scenic spot, at one end of the large oval lawn. The lawn is cut into irregular triangles or trapezoids like a cake by a grid of paths. Around the building, the paths sink, and the buildings and flower pools stand out as a set of high and low cut forms that become the center of the large lawn, the artwork of the art district.

建筑是用来表演的

如果说游客服务中心是有功能的，那么除了为游客提供信息服务、咖啡茶饮、休憩盥洗外，还有一个因场地需要而产生的表演功能。由于建筑在椭圆形大草坪的一端，另一端地形略微隆起，草坡上铺以条石，自然成为室外看台，而建筑呼应看台就成了舞台。我们把下沉建筑的屋顶作为舞台正好满足了这种需求，除了专业团体，游客也可以在此自由舞动，享受看与被看的乐趣。

建筑是用来体验的

这个建筑是有趣的，因为它是可以体验的，与一般建筑注重室内空间不同，这种体验是沿着建筑外侧穿行时才能感觉到的，而且是与高起的景观花池一同体验的。沿着分不清是建筑边界还是花池侧壁所限定的下沉小路穿梭，两侧高高低低的红砂岩碎拼墙体使得视线时而开放，时而封闭，人们很容易在不同的空间体验中找到乐趣。

建筑是因景随宜的

景观建筑某种程度上可以理解为景观中的构筑物，比如亭廊台榭等，所不同的是它更能挡风遮雨，适合不同的季节冷暖。景观中的建筑，建筑属性弱化，景观属性增强，到一定程度，二者之间的边界就变得模糊起来。也就是说，景观中的建筑与一般意义上的建筑是不同的，它是随环境而变的。正如《园冶》中“屋宇篇”所示：“凡家宅住房，五间三间，循次第而造；惟园林书屋，一室半室，按时景为精。方向随宜，鸠工合见；家居必论，野筑惟因。”

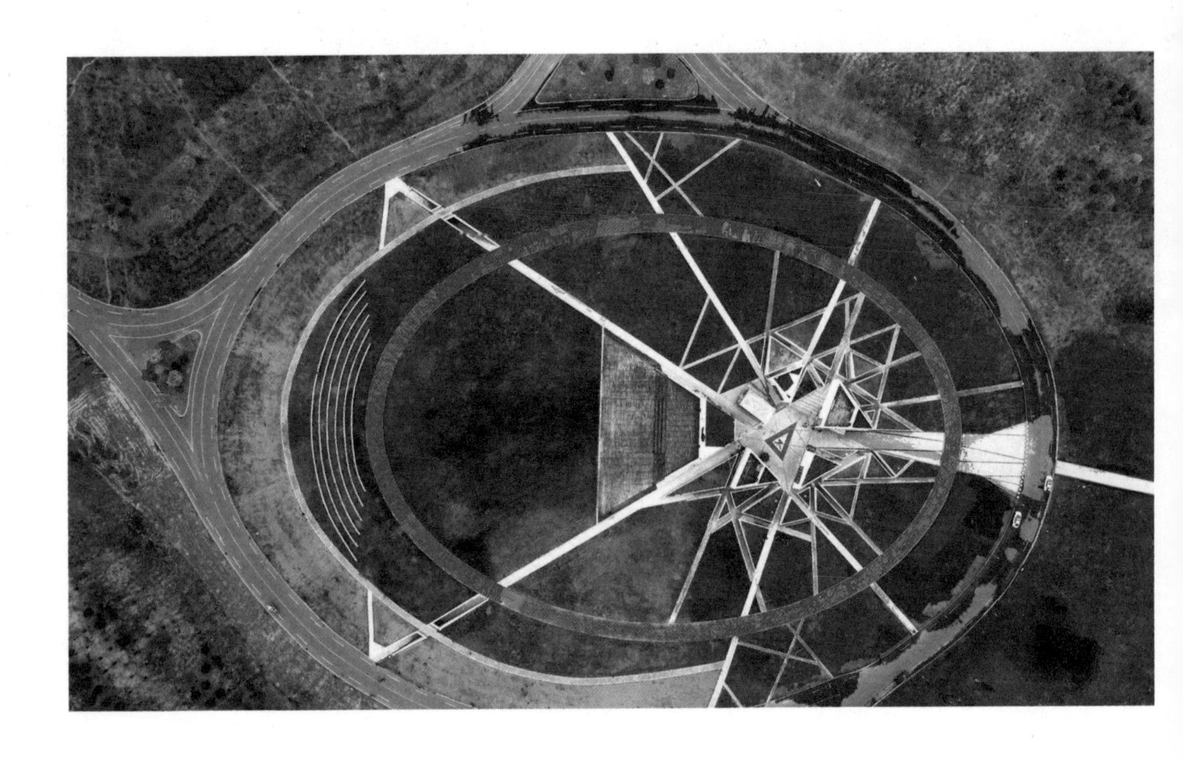

Architecture Is For Performing

If the tourist service center is functional, then in addition to providing tourists with information services, coffee and tea, rest and toilet, there is also a performance function generated by the needs of the site. As the building is located at one end of the oval lawn and the terrain at the other end is slightly raised, the grassy slope is paved with strips of stone, which naturally becomes the outdoor grandstand, and the building echoes the grandstand and becomes the stage. We use the roof of the sunken building as a stage to meet this need. In addition to professional groups, visitors can also dance freely here and enjoy the pleasure of seeing and being seen.

Architecture Is For Experiencing

This building is interesting because it can be experienced.Different from the general buildings focused on interior space, this experience only can be felt when walking along the outside of the building, and is experienced in conjunction with the elevated landscape flower pool. Shuttle back and forth along the sinking path which is unclear whether it is the building boundary or defined by the side wall of the flower pool. The high and low red sandstone patchwork walls on both sides make the view open and closed at times, making it easy for people to find pleasure in different spatial experiences.

Architecture Is Appropriate for the Scene

To some extent, landscape architecture can be understood as the structures in the landscape, such as pavilions, corridors, terraces, etc. The difference is that it can better shelter from the wind and rain, suitable for different seasons. Architecture in the landscape, the architectural attribute is weakened, the landscape attribute is enhanced, to a certain extent, the boundary between the two becomes blurred. In other words, the architecture in the landscape is different from the architecture in the general sense, and it changes with the environment. As is shown in the building chapter in "Yuan Ye" :"The residential building that every household lives in, no matter 3 rooms or 5 rooms, must be built according to dimensional sequence successively.Only garden buildings, whether one room or half a room, should be built according to the changing landscape of the seasons. It is the consensus of gardeners that landscape architecture is not restricted by orientation.Residential building should pay attention to the rules and regulations, while garden building should be specific to local conditions."

首都博物馆

Capital Museum

首都博物馆成立于1981年10月。在过去近20年的时间里，它一直静静地“隐居”在京城国子监孔庙里。为了扩大首都博物馆的影响，提高城市的文化品位，北京市政府和国家文物局决定在西长安街延长线白云路的西侧建设首都博物馆新馆。区位敏感、用地紧张是地块的优势与不足，而建筑体量大、管井密集是景观设计的挑战，也就是如何在有限的空间中解决各类矛盾，使景观与建筑融合，形成一体化效果。

The Capital Museum was established in October 1981. For nearly past 20 years, it has quietly been "a hermit" in the Confucius Temple of the Imperial College in the capital. In order to expand the influence of the Capital Museum, and improve the cultural taste of the city, the Beijing Municipal Government and the State Administration of Cultural Heritage decided to build the new Capital Museum on the west side of Baiyun Road in the extension of west Chang 'an Avenue. The advantages and disadvantages of the plot are sensitive location and tight land use, while the large building volume and dense tube well are the challenges of landscape design, that is, how to solve all kinds of conflicts in limited space, so as to integrate landscape and architecture and form an integrated effect.

引申建筑理念

在充分理解建筑设计理念后，景观首先是要“开放”，其次是要“意境”。开放就是要突出建筑前的广场意象，既要与长安街整体街景协调，又要突出建筑的宏伟庄重和开放亲民。意境就是要让建筑周边景观充满文化意境和人文情怀，与博物馆气质相协调。

Extend the Concept of Architecture

After fully understanding the concept of architectural design, the landscape should be "open" first and "artistic conception" second. Open is to highlight the image of the square in front of the building, not only to coordinate with the overall street scene of Chang 'an Avenue, but also to highlight the building's magnificent, solemn and amiable. Artistic conception is to make the surrounding landscape of the building full of cultural conception and humanistic feelings, in harmony with the temperament of the museum.

延续建筑肌理

建筑北立面的构图充满理性与浪漫，富有地标性特点。与此对应的北广场在构图上延续了建筑柱网模块肌理，形成规则均衡的井田式布局，主入口的形式、材料从室内向外延伸，铺装材料的大小、颜色、质感与建筑立面保持一致，形成建筑景观一体化效果。建筑东侧从东北角向西南角缓缓下沉延伸至室内空间的楔形院落，种植竹林，形成室内外的一体化联通。

提炼建筑语言

建筑的空间语言和材料语言丰富多彩，景观设计要提炼建筑语言并巧妙处理细部与建筑融合。“水院”利用兽头喷水、砖雕、木格栅、玻璃、陶瓷雕塑等元素营造了“高墙水院”庭院气氛，既改善了新风采集和安防功能，又使建筑有一种从地下升起的感觉。“竹院”将城市景观引入建筑内部，打破了传统博物馆封闭沉闷的感觉，为市民营造了明亮温馨、开放亲民的景观意境。

Continue the Architecture Texture

The composition of the north facade of the building is rational and romantic, full of landmark characteristics. The corresponding north square continues the module texture of building column grid in the composition, forming a regular and balanced layout of the well field system. The form and materials of the main entrance extend from the interior to the outside, and the size, color and texture of the paving materials are consistent with the building facade, forming an integrated effect of architectural landscape. The east side of the building descends from the northeast corner to the southwest corner and extends into the wedge-shaped courtyard of the interior space, where bamboo forests are planted to form an integrated connection between interior and exterior.

Refine Architectural Language

The spatial language and material language of architecture are rich and colorful, and the landscape design should refine the architectural language and skillfully integrate the details with the architecture. "Water courtyard" uses the elements such as water spraying from animal head, brick carving, wood grilles, glass and ceramic sculpture to create the courtyard atmosphere of "high wall and water courtyard", which not only improves the function of collection of fresh air and security, but also gives the building a feeling of rising from underground. The "bamboo courtyard" introduces the urban landscape into the interior of the building, breaking the closed and depressing feeling of the traditional museum, and creating a bright, warm, open and intimate landscape mood for the public.

3. 相辅相成

Complement Each Other

外研社大兴国际会议中心

Daxing International Conference Center of Foreign Language Teaching and Research Press

施工中的场地

2004年除夕的前一天，我们驱车大兴，时值隆冬，外研社大兴国际会议中心的工地冷冷清清，地上堆满了建筑渣土、钢筋水泥和模板等施工物料。寒风穿过建筑间隙，在塔吊和脚手架上发出“嘶嘶”的声音。在这种声音的伴随下，我们爬上爬下地走完了场地的每一个角落，努力寻找着感觉，想象着春暖花开后的场景。踏勘结果：带回70Mb的照片和对五一前全面完工的疑虑，轮胎被扎两个，略感风寒。

整合的过程

之后的三个月，整合景观与建筑的关系成为主线。与李社长碰方案、与崔总定方案、与建筑师沟通、画图、改图、出图、日日夜夜加班加点。施工过程中数次的现场配合，与施工单位解决各种问题、与康体设施公司确定水池结构及设备方案、与石材厂家选石定料、与植物供应商看苗圃选苗木……一切一切历历在目。

Site Under Construction

On the day before the Spring Festival's eve of 2004, we drove to Daxing in the middle of winter. The construction site of the Daxing International Conference Center of Foreign Language Teaching and Research Press was desolate. The ground was filled with construction materials such as construction waste residue, steel and cement, and form boards. Wind chilled sizzle through gaps in buildings, hissing on tower cranes and scaffolding. Accompanied by this sound, we climbed up and down to every corner of the site, trying to find the feeling and imagine the scene after the spring blossoms. Survey results: brought back 70mb of photos, doubts about the full completion before May 1, and the two tires were pricked, slightly caught a cold.

Process of Integration

Over the next three months, the integration of landscape and architecture became the main thread. We met the plan with President Li, made the plan with President Cui, communicated with the architect, drew, modified and produced the plan, and worked overtime day and night. During the construction process, we cooperated with the construction unit on site several times to solve various problems, determined the structure of the pool and equipment scheme with the Kangti facilities company, selected stone materials with the stone manufacturer, and looked through the nursery garden and selected the nursery stock with plant suppliers··· Everything is in our mind.

与建筑对话的植物

场地中的植物配置有两个特点，一是竹子的选用，二是树阵的形成。几个独立的小餐厅和游泳馆外大量种竹，除了营造氛围、遮挡视线外，夜间在室内灯光的映衬下，颇有“墨竹”的神韵。在庭院的不同位置，结合不同场景配置不同植物的树阵。它们分别是银杏阵、玉兰阵、七叶树阵、马褂木阵、栾树阵。相同植物构成的树阵形成了植物的韵律及内在逻辑，与建筑形成了良好的对话。

呼应建筑的休息亭

在培训楼室外的公共通路上有5个休息亭，对应每一个出入口。休息亭是混凝土结构，顶、墙、地一体化形成“C”字形态，虽然形态完全一致，但5个亭子贴面材料各不相同，分别为红砖、灰砖、青色页岩、白色面砖、黑色马赛克，呈现出不同的颜色和肌理，也给场地增加了情趣。休息亭的墙面上嵌有座椅、电话、告示牌和垃圾箱，体现了对室外活动的人性关怀。

串联室内外的温泉池

室外的5个温泉池为整体环境增色不少。为了防止池壁过厚比例不当，设计中与设备安装公司协商，将进出水管埋于混凝土池壁内，增加了池子外的活动场地范围。池壁采用石材马赛克和水晶马赛克，北侧封闭小院内两个温泉池一黑一白，颇有禅意。南侧三个池子选用红蓝绿三种颜色，与红墙、蓝天、绿地相呼应。水池周边尝试了枯山水的造园手法，不仅用于内庭院而且还在草坪上加以发挥，形成类似高尔夫果岭的效果，增加了趣味性。

属于建筑的景观故事

正如崔总在工程总结会上所言“每一个建筑都有属于它自己的故事”。这些故事源自于熟悉它的人，在喜欢它的人中间流传。对于这座建筑群而言，我们所有的回忆与记述都成为故事的一个个片断，在故事的片断中，我们感到欣慰、得到经验、找到启发。看着如今草长莺飞的场景，回想当初的遍地瓦砾，真感时间的伟大和人工的神奇。作为景观建筑师，我们与建筑的故事还在继续，愿我们能以自己的心血和热情，不断书写出更多属于建筑的景观故事。

Plants That Speak to Architecture

The plant configuration in the site has two characteristics, one is the selection of bamboo, the other is the formation of tree array. A large number of bamboo were planted outside a few independent small restaurants and swimming pool. In addition to creating an atmosphere, blocking the line of sight, under the set off of the indoor lights at night, there is a verve of "ink painting of bamboo". At different locations in the courtyard, the tree arrays of different plants are arranged in combination with different scenes. They are ginkgo biloba array, yulan array, buckeye array, liriodendron chinese array, goldenrain tree array. The tree array formed by the same plant forms the rhythm and internal logic of the plant, forming a good dialogue with the building.

Rest Pavilions That Echoes the Architecture

There are five rest pavilions on the public access road outside the training building, corresponding to each entrance and exit. The rest pavilion is a concrete structure, the top, wall and the ground are integrated to form the "C" shape. Although the form is exactly the same, the five pavilions are covered with different materials, including red brick, ash brick, cyan shale, white face brick and black mosaic, presenting different colors and textures, which also add interest to the site. The wall of the rest pavilion is embedded with seats, telephones, signs and dustbins, reflecting the human concern for outdoor activities.

Hot Spring Pools That Connect Indoor and Outdoor

Five outdoor hot spring pools grace the whole environment. In order to prevent the improper proportion of excessive thickness of the pool wall, during the design, we negotiated with the equipment installation company to bury the inlet and outlet pipes in the concrete pool wall, which increased the scope of activity outside the pool. The wall of the pool is made of stone mosaic and crystal mosaic. There are two hot spring pools in the closed yard of the north side, one black and one white, with quite a zen. The three pools in the south are red, blue and green, which echo with the red wall, blue sky and green space. Around the pool, we tried the gardening technique of dry landscape, which is not only used in the inner courtyard but also played on the lawn, creating an effect similar to the golf green and increasing the interest.

Landscape Story That Belongs to Architecture

"Every architecture has its own story," as President Cui said at the project summary meeting. These stories come from people who are familiar with it, and they circulate among people who like it. For this group of architectures, all our memories and descriptions become fragments of the story, in which we are gratified, gain experience, and find inspiration. Looking at the scene that grasses are tall and the nightingales are in the air, thinking back to the original rubble all around, we truly feel the greatness of time and artificial magic. As landscape architects, the story between us and architecture is still going on. We hope we can write more landscape stories that belong to architecture with our own efforts and passion.

百度科技园
Baidu Science Park

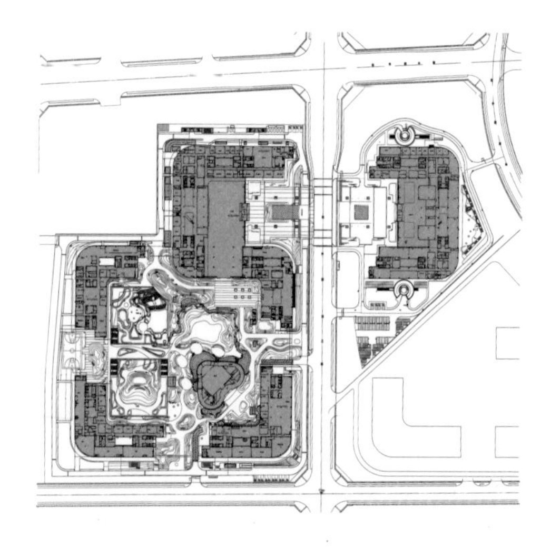

科技与诗意

百度缘起“梦里寻她千百度，蓦然回首，那人却在灯火阑珊处”，“科技搜索”与“诗意远方”本就是理性与感性的互补和碰撞。百度因向往而搜寻，因梦想远方的心而诗意。“搜索引擎框”外形的建筑群构建出了外向自然、内向庭院的空间格局，心中众人千里，院中一树一山一湖，合了百度翘楚之欲又兼顾了诗意情怀。园林由于诗意而定格在“虽由人作，宛自天开”的意境之中。

因形与达意

若有象形，若无有意。建筑的南北两区因道路相隔，分离与阻断势必引发南北同体的愿望。所以在南北两个“C”形广场上设置了水面与草坡，借助空间及形态上的一致形成南北同体的感觉。院内一共有5处水景，虽是独立的系统，形态上却互为关联。水流高来低去的势态和形态变化使人产生联想，形生意，意补了形，而心随意动。

Technology and Poetry

Baidu derived from: "I looked for her thousands of times in the dream, suddenly turned around, but saw her standing in the dim light", "scientific and technological search" and "poetic distance" is the complementation and collision between reason and perceptual.Baidu searches because of yearn, and becomes poetic because of the heart dreaming of the distance. The architectural complex with appearance of "search engine box" builds the spatial pattern of extroversion nature, introversion courtyard. There are thousands of people and thousands of miles in the heart, and a tree, a mountain and a lake in the courtyard, which fits the desire of becoming outstanding with poetic feelings of Baidu. The garden is frozen in the artistic conception of "although people do it, it as if is made by nature" because of the poetry.

Form and Meaning

If there are, there are shapes. If not, it has meaning. The north and south areas of the building are separated by roads, and the separation and block will inevitably lead to the desire of the combination of north and south. Therefore, water surface and grassy slope are set on "C" type squares in the north and south side to form a sense of north and south identity with the help of the consistency of space and form. There are altogether 5 waterscapes in the courtyard, although they are independent systems, they are related to each other in form. The state and form changes of water flow coming from high and leaving in low make people think. The shape generates the mind, the mind complements the shape, and the feeling changes with mind.

敞阔与幽秘

内向庭院分为上院和下院，下院又名“夏院”，突出盛夏的浓荫；上院以展现春华秋实为主题。建筑设计的格局无论上、下院子都是敞阔的，为了形成感受差异，景观处理上上院更为敞阔，下院特意幽秘。这是因为上院的土层厚度受限，大树难以存活，于是塑造“一山一湖”，周边以微地形增加种植土厚度，丰富植被层次。土层薄，场地多，便利了行人的停停走走，流动的风景便在上院形成。讨巧的是上院绿意浓荫的部分是因借了下院的大树冠稍。下院的树是养在实土上的，尤其地繁茂，绿色毫不吝惜地给了上院，也成了人们愿意停留的去处。下院的好补了上院的缺，上院的来往生机补了下院的幽秘静谧，园内形成了浑然一体又迥乎相异的两个院落。

Open and Secluded

The inner courtyard is divided into the upper house and the lower house, and the lower house is also known as "summer courtyard", highlighting the shade of high summer; the theme of the upper house is blossoms in spring and bears fruit in autumn. In the pattern of architectural design, both the upper and lower houses are open and wide. In order to form a feeling of difference, the upper house is opener and wider in landscape treatment, while the lower house is deliberately secluded. This is because the thickness of the upper house soil layer is limited, and the trees are difficult to survive, so "one mountain and one lake" is created, and the surrounding area is enriched with micro-topography to increase the thickness of planting soil and enrich vegetation layers. There are thin soil layer and many sites, which facilitates the stop and go of pedestrians, so the flowing scenery is in the upper house. It is fortunate that the leafy part of the upper house is borrowing the large canopy of the lower house. The trees in the lower house are kept on solid soil, especially luxuriant. The green leaves are given to the upper house without stint, and they also have become a popular place for people to stay. The advantages of lower house make up for the disadvantages of the upper house, and the vitality of the upper house makes up for the privacy of the lower house. There are two courtyards formed in the garden, a unified entity but different with each other.

二　人与自然和谐共生

Harmony Coexistence Between Man and Nature

第十二届中国（南宁）国际园林博览会

The 12th China (Nanning) International Garden Expo

知山知水

南宁是以邕江河谷为中心的盆地区域。这个盆地向东开口，南、北、西三面均为三地围绕，北为高峰岭，南有丘陵，西有凤凰山，形成了西起凤凰山，东至青秀山的长形河谷盆地。南宁地形多样，有平原、盆地、丘陵、山地，以平原和丘陵为主。南宁园博园处于南宁地形中的丘陵地带，现状生态资源得天独厚，用地内山岭纵横、江水蜿蜒、林木茂盛、花田沃野、湖塘共鸣，山、水、林、田、湖、草生命共同体系统完整、特征明显。

Know the Mountain and Water

Nanning is the basin area that centers on the Yongjiang river valley. This basin opens to the east, and south, north and west are surrounded by three sides. The north is Gaofeng Hill, south has the hill, west has the Phoenix Mountain, which formed a long valley basin, west from the Phoenix Mountain, east to the Qingxiu Mountain. The terrain type of Nanning is diverse, there are plains, basins, hills, mountains, mainly plains and hills. Nanning Garden Expo is located in the hilly terrain of Nanning, with unique ecological resources. The land is full of crisscrossing mountains, winding rivers, lush trees, fertile fields and ponds, and the community system of mountain, water, forest, field, lake and grass life is complete with obvious characteristics.

Yudong Road
Zhonglong Road
Puxing Road
0 100 200 500m
1 Exquisite Lake
2 Linglong Island
3 Waterfront Teahouse
4 China-ASEAN Loggia
5 Southeast Asia Botanical Garden
6 Shopping Street
7 Tourist Center
8 Arhat -pine Court
9 Crystal Spring Lake
10 Crystal Spring Tower
11 Floral Brookvale
12 Performance Center
13 Clear Spring
14 Livable Cities Gallery
15 Kid's Wonderland
16 Management Center of Park
17 Heart-shape Harbour
18 Horse Pasture
19 Dingsishan Mound Museum
20 Dingsishan Mound Park
21 Rushes Water Retreat
22 City Gardens
23 Parking Lot
2016.10
2016.12
2018.02
2018.03
2018.04
2018.06
2018.09
2018.12

轻拿轻放

如此特征的现状条件，必不能按照惯常的规划设计手法，一切皆宜轻拿轻放。规划设计凭借山、水、林、田、湖、草等得天独厚的自然环境以及丰富多彩的民族文化和面向东盟的区位优势，以“生态宜居，园林圆梦”为主题，按照“特色南宁铸就不一样的园博园”的规划目标，在充分尊重现状地形地貌、山形水系的基础上，本着“不推山、不填湖、不砍树”的理念,充分利用设计智慧将南宁园博园所有设计轻轻地放入自然山水之中，形成特色的山水意象。

Handle With Care

The status quo conditions with such characteristics must not be in accordance with the usual planning and design techniques, and everything should be taken lightly and handle with care. With mountains, water, forests, fields, lakes and grasses, advantaged natural environment, rich and colorful national culture and regional advantages facing ASEAN, the planning and design takes "pleasant living environment, fulfilling dream in gardens" as the theme, and follows the planning goal of "creating a garden expo garden with distinctive characteristics in Nanning". On the basis of fully respecting the current situation of topography, landform mountain shape and water system, in line with the concept of "don't push mountains, don't fill lakes and don't cut trees", we make full use of the design wisdom to gently put all the designs of Nanning Garden Expo into the natural mountains and rivers, forming a characteristic image of mountains and rivers.

优景优境

著名科学家钱学森先生曾在给吴良镛院士的信中写道：“山水作为人们寄托理想、慰藉心灵和调节身心的载体，与城市互为补充，体现了人们对城市生活与乡野生活的共同需求。若将自然中的大山大水和城市联系起来进行现代城市总体格局的把握则会构建凸显地方山水特色的人类理想居住环境。”对南宁而言，园博园是城市山水系统中的一个节点，其山水脉络要和周边，甚至整个城市的山水相连，形成大系统大格局。对园博园而言，因其内部就有众多山水丘陵，只要通过少量的艺术化人工干预，就能打造出一个完整的山水系统，来凸显园博园的山水魅力。“三湖六桥十八岭”就是在充分尊重现状地形地貌的基础上，通过适当的设计梳理和艺术加工，呈现出园博园“山水奇”的特色。而依据园博园现状景观资源及地形地貌的特点，因地制宜、依景造境，规划了芦草叠塘、玲珑揽翠、松鼓迎宾、花阁映日、清泉明月、潭池寄情、矿坑七彩、贝丘遗风八大景观，步移景异、引人入胜。

Good Landscape and Good Environment

Qian Xuesen, a famous scientist, wrote in a letter to academician Wu Liangyong, "Mountains and rivers, as a carrier for people to place their ideals, comfort their soul and regulate their body and mind, is complementary to the city and reflects people's common needs for urban life and rural life. If the mountains and rivers in nature are connected with the city to grasp the overall pattern of the modern city, an ideal living environment highlighting the local characteristics of mountains and rivers will be built." For Nanning, the garden expo is a node in the system of mountains and rivers of the city. The landscape context should be connected with the surrounding area, or even the whole city, to form a large system and pattern. For the garden expo, because its interior has numerous mountains, rivers and hill, as long as there is a small amount of artistic manual intervention, a complete system of mountains and rivers can be created to highlight the charm of mountains and rivers of the garden expo. "Three lakes, six bridges and eighteen ridges" based on the full respect of the current topography and landform, through appropriate design combing and artistic processing, presents the characteristics of "amazed mountains and rivers" of the garden expo. According to the landscape resources of the expo garden and the characteristics of the terrain and landform, we adjust measures to local conditions and create the environment according to the landscape. We planned eight landscapes of reed grass stacked pond, Linglong Lancui, pine and drum welcoming guests, flower pavilion reflecting the sun, clear spring and bright moon, pool reposing emotion, mine pit with seven colors, and shell mound in relics style. Different landscape changes as moving, leading people into the beautiful scenery.

心象自然

南宁园博园的规划设计采用“心象自然”的设计理念，因自然而得心象，借自然而成心象。从规划前期开始，规划设计团队借助多种技术，多角度、多层次地进行广泛深入的现场调研，深入培养设计师对场地的感情，使其对场地中的山水林田湖草充满无限激情和敬畏之心。随着调研的深入，一幅自然隐逸的林泉图景便在设计师心中逐渐清晰起来，而这幅图景在潜移默化地引领着设计师小心翼翼地将规划设计轻轻地放入园博园这片自然山水之中。“一阁四馆两中心”等建筑被轻轻地放入山体中，园路广场被轻轻地放入山脚下，城市展园被轻轻地放入山坡上，雨水花园被轻轻地放入鱼塘中等等。所有的设计好像原本就应该是这样生长在这里似的，既顺应了自然，又呈现了内心的意象，所谓“象为自然，与心相应”。正如李渔所说“山水者，情怀也；情怀者，心中之山水也。”

Nature in Mind

The planning and design of Nanning Garden Expo adopts the design concept of "nature in mind", which can gain mind by nature and create mind by nature. From the early stage of planning, the planning and design team use a variety of technologies, conduct extensive and in-depth field research from multiple perspectives and at multiple levels, deeply cultivate the designer's feelings for the site, and have full of infinite passion and awe for the mountains, rivers, forest, land and grass in the site. As the research goes deeper, a forest-spring landscape of natural seclusion gradually becomes clear in every designer's mind, and this landscape unconsciously leads the designer to carefully put the planning and design into the natural mountains and rivers of the garden expo. "One pavilion, four houses and two centers" were lightly put into the mountain, the garden road square was lightly put into the foot of the mountain, the city exhibition garden was lightly put into the hillside, the rain garden was lightly put into the fish pond and so on. All the designs seem to be originally supposed to grow here like this, which not only conform to the nature, but also present the image in the heart, so-called "the image is the nature, corresponding with the heart". As Liyu said "Mountains and rivers, are the feelings; Feelings, are the mountains and rivers in the heart."

 Inner Nature Imagery

2019北京（延庆）世界园林艺术博览会

2019 Beijing (Yanqing) World Garden Art Expo

心仪
世园中的自然存在

2014年的晚春，我们踏入了位于延庆的北京世园会的选址之地，这里是毗邻八达岭长城、远望海坨山，横跨妫河两岸，东临延庆新城的生态涵养之所。妫水河畔芦苇荡漾，鸟鹭齐飞；妫河森林公园林木葱郁、虫蛙齐鸣；广阔的田地间、村落旁星星点点地大树点缀其间，这里就是2019年世园会所在地原来的样子。

Admire in the Heart — Natural Existence in the World Garden Art Expo

In the late spring of 2014, we stepped into the site of the Beijing World Garden Art Expo in Yanqing, which is the ecological conservation place, adjacent to the Badaling Great Wall, overlooking the Haituo Mountain, across the Gui River and east of the new district of Yanqing. Reeds rippled near the Guishui River with birds and herons flying; lush trees were verdant in the Guishui Forest Park with insects and frogs singing; vast fields and villages were dotted with trees. This is the original look of site of the 2019 world garden art expo.

保留
世园中的林荫瑰宝

“要保护这里的每一条河流，保留每一片大树，因为它是园子里自然生命的见证，是园子里最宝贵的财富！”在景观设计之初，我们就立下了保留现状树的军令状，这意味着从全局意义上而言“保留大树”成了设计的重点。

园子中最难保留的要属2号门区的大杨树。2号门是园区最主要的门区之一，除了形象需求外，还有较大的集散及通行功能需求，但由于市政道路标高较高，且门区建筑正负零标高需满足城市50年一遇防洪水位标高要求，致使入口广场的标高需要整体抬升，与其有效衔接，如果整体抬升，那么位于门区西南角的高度8～12米的现状大杨树会被广场掩埋0.6～1.4米，怎么办？最终我们为大杨树们设计了一个低洼的雨水花园，既保留了原生场地的标高关系，又为大树预留了生长空间。

Reserve — the Tree-lined Treasures of the World Garden

"To protect every river here, to preserve every tree, because it is the witness of natural life in the garden, is the most valuable wealth in the garden!" At the beginning of the landscape design, a pledge was established to preserve the existing trees, which means that "preserving the trees" has become the focus of the design from a global perspective.

The tree hardest to reserve in the garden is the big poplar in gate no.2. Gate no. 2 is one of the most important gate areas in the park. In addition to image requirements, there are also large demand for gathering and distributing and traffic functions. However, because the elevation of municipal roads is higher, and the plus or minus zero elevation of buildings in gate area should meet the elevation requirement of flood control level that cities come once in 50 years. As a result, the elevation of the entrance plaza needs to be raised as a whole to effectively connect with it. If the whole area goes up, the existing poplar trees at the southwest corner of gate area with a height of about 8-12 meters will be buried by the square about 0.6-1.4 meters. What's to be done? In the end, we designed a low-lying rain garden for the poplars, preserving the elevation of the original site while leaving room for the trees to grow.

园子中树龄最大的要属园区东侧的两排大旱柳，大家亲切地叫他们柳树伯伯。《诗经·小雅·采薇》中的“昔我往矣，杨柳依依”，描绘的就是“难分难离，不忍相别，恋恋不舍的心意”，设计中就是带着这样的心境留存了这一片美好。沿着国际展园中央轴线的水溪向东北望，保留的大柳树穿插在体验馆间与建筑相互掩映，生动地描画出“柳暗花明又一村”的田园景象，也为园子保留了春天里最早的故事。

园子中最舒适的地方当属北部的一大片原生刺槐林，那是场地中历史自然的印记，同时也是园区中一大片林荫空间。我们毫不犹豫地保留了刺槐林，并量身定做了绕行的小尺度生态园路，同时利用北部现状低洼地设计了一条雨水花溪与一片疏林草地与其作伴。夕阳西下，温暖的阳光投射其间，我们穿行在林荫之中，望见雨水花园与刺槐林安静地相依。

共生
世园中的蝴蝶与花

蝴蝶会恋上花，花也会爱着蝴蝶，也许这就是上天的造化，也许是园艺中自由、美丽的故事。世界园艺轴作为园区主要轴线之一，串联着国际馆及演艺中心两个核心建筑。国际馆由94朵钢铁“花伞”组成，如同一片花海飘落在园区里，演艺中心宛如一只彩蝶驻足在妫湖之滨，轴线景观就在“花”与“蝴蝶”的故事中飞舞共生。“穿花蛱蝶深深见”出自唐代诗人杜甫的《曲江二首　其二》，描绘的是蝴蝶在花丛深处穿梭往来、时隐时现的场景。轴线就是以其为设计原型，以花开蝶舞为主题，引种各国花卉，提取蝴蝶元素，打造绚丽的国际园艺景观，形成大自然动植物和谐共生的景象。

The oldest tree in the garden belongs to two rows of dry-land willows on the east side of the garden, which were affectionately called uncle willow. In the book "book of songs. Xiaoya. Caiwei", there is "recall the original when I went out to battle, the willows swayed in the wind." It describes the feeling that is difficult to separate and say goodbye and is reluctant to part from. Design is with such a state of mind retained this piece of beauty. Look towards northeast along the stream which is the central axis of the international exhibition garden, the preserved willows are interspersed into the experience pavilion and set off with the architecture. It vividly paints a rural scene that is "willows are dim, flowers are bright and suddenly a mountain village appeared before the eyes" , but also keeps the earliest stories in the spring for the garden.

The most comfortable part of the garden is a large area of native locust forest in the north, which is a mark of the historical nature of the site and a large shade space in the park. We did not hesitate to retain the locust forest, and customized a circuitous ecological garden road in small scale. At the same time, a rainwater flower-stream and a patch of forest and grass were designed based on the low-lying land in the north for company. The warm sun is setting down and the sunshine is shining as we walked through the trees, looking at the quiet dependence of the rainwater gardens and locust forest.

Symbiosis — Butterflies and Flowers in the World Garden

The butterfly will fall in love with the flower, the flower will also love the butterfly, maybe this is the creation of heaven, maybe it is the free and beautiful story in gardening. As one of the main axes of the garden, the world horticulture axis connects the two core buildings of the international pavilion and the performing arts center. The international pavilion is made up of 94 steel "flower umbrellas", which are like a sea of flowers falling down in the park. The performance center is like a colorful butterfly standing on the Gui Lake, and the axis landscape is flying and dancing in the story of "flowers" and "butterflies". "Butterflies shuttled through the depths of the flowers" is from the "Qujiangershou-Qier" by Dufu in Tang Dynasty, depicting a scene that the butterflies shuttling back and forth in the depths of the flowers, appeared and disappeared. The axis is to take it as the design prototype, take flower blooming butterfly dancing as the theme, introduce flowers from various countries, extract butterfly elements, create a gorgeous international horticultural landscape, and organize the scene of harmonious coexistence of natural plants and animals.

超能
世园中的五大精灵

园艺承载着人类对自然的向往融入生活的各个角落。在园区体验带及国际展园公共区，我们将与植物生长息息相关的“风、光、水、土、温度”这五大精灵融入设计，将游客的视觉、听觉、嗅觉、味觉、触觉融入园艺体验，注重景观与人的互动性，以交互式体验设施增强人与自然的接触度。

种子生长基质——土壤因子，种子的足迹由此展开，开始生根萌发。
种子传播媒介——风因子，风媒作为种子的传播方式之一，为种子的足迹提供无限可能。
种子生长条件——阳光因子，种子破土之后，在阳光滋润下茁壮成长，完成生命周期。
种子生长之源——水因子，种子在水分的滋润下破土而出，水滋润萌芽继续生长。

种子生长要素——温度因子，种子的生长和繁殖要在一定的温度范围内进行，在最适温度范围内植物生长繁殖得最好。不同种子在不同温度的影响下，生长状况也呈现出多样性。

同时，对于展园围合的公共区景观注重游客使用需求，以人为本，提供充足且具有特色的林下休憩空间及配套服务设施，形成聚集人气的公共休憩场所。

Super — the Five Spirits in the World Garden

Gardening carries human's yearning for nature into every corner of life. In the experience zone of the park and the public area of the international exhibition park, we integrate the five spirits of "wind, light, water, earth and temperature" closely related to plant growth into the design, integrate the visitor's vision, hearing, smell, taste and touch into the horticultural experience. The interactivity between landscape and human is emphasized to enhance the contact between human and nature with interactive experience facilities.

Seed growth matrix — edaphic factor, spreads the seed's footprint and begins to take root.
Seed media — wind factor, as one of the mode of transmission of seeds, wind media provides unlimited possibility for seed footprint.
Seed growth condition — sun factor, after breaking through the soil, the seeds thrive under the moisten of the sun, complete the life cycle.
Seed growth source — water factor, the seeds in the moisten of water break through the soil, the buds continue to grow under the water moistens.

Seed growth element — temperature factor, growth and reproduction of seeds should take place within a certain temperature range, and plants grow best in the optimum temperature range. Under the influence of different temperature, the growth condition of different seeds also shows diversity.

Pay attention to the needs of tourists for the public area landscape enclosed by the exhibition garden at the same time. To be people-oriented, and provide sufficient and distinctive rest space under the trees and supporting service facilities to form a popular public rest place.

生长
大地中的中国馆

“设计始于对园艺、农耕和自然的理解——园艺脱胎于农耕，梯田是农耕文明的独特景观、人与自然和谐共生的典范”，中国馆景观在崔愷院士提出的理念下展开。

创源——以梯田为缘起，在高低起落的坡地上，设计出或宽或窄的梯田，不规则地依山势上下伸展，生长于妫汭湖畔，以半圆形轮廓融入场地，将中国馆舒展优美的曲线屋顶包裹其间，诉说着园艺起源于农耕文明的故事。

造景——梯田运用生态石笼挡墙，建立棱角分明，梯级层次清晰的台地系统，内填延庆本地石料，用于山体结构的稳固与美观，蜿蜒的石笼墙如同长城绵延，为“长城脚下的世园会”点睛。在梯田外围种植不同植被品种调节环境的温度、湿度与色彩，同时通过植物根系与土壤的联系，增强水土保持效应，建立一个自我循环的生态环境。在中国馆前广场中央核心，指导农业生产和生活的二十四节气与下沉水院交互成趣，向世人展示“中国的第五大发明”。

达意——回望中国馆，层叠的梯田，葱茏的树荫、多彩的植被与建筑完美交融，人随梯田逐级而上，在田埂行走间完成园艺精神的升华，如同一部刻录着中国农耕文明成果与园艺生活的立体史册，与世园胜景交相辉映，默默守望着我们的家园。

Growth — China Pavilion in the Earth

"Design begins with an understanding of gardening, farming and nature — horticulture grew out of farming, and terraced fields are the unique landscape of farming civilization and a model of harmonious coexistence between man and nature." The landscape of the China Pavilion unfolds under the concept proposed by academician Cui.

Create the source — derived from terraced fields, on the high and low slopes, wide or narrow terraced fields were designed to extend irregularly up and down according to the mountain situation. Terraced fields grow in side of Guirui Lake, blend into the site with a semicircular outline, enclose the extended and graceful curvilinear roof of the China Pavilion, tell the story of gardening originated in the farming civilization.

Create the landscape — the terrace fields use the ecological stone cage to block the wall, establish a platform system with sharp angles and clear levels, fill in the local stones of Yanqing for the stability and beauty of mountain structure. Winding stone cage wall is as the Great Wall stretches, providing striking key points for the "world expo at the foot of the Great Wall". Different vegetation varieties are planted in the periphery of the terrace to regulate the temperature, humidity and color of the environment. Meanwhile, through the connection between plant roots and soil, soil and water conservation effects are enhanced to establish a self-circulation ecological environment. In the central core of the square in front of the China Pavilion, the 24 solar terms that guide agricultural production and life are interacted with the sunken water courtyard, presenting "China's fifth great invention" to the world.

Express the artistic conception — looking back at the China Pavilion, the cascade terrace, verdant shade, colorful vegetation and architecture blend perfectly. People go up step by step with the terrace, walking in the ridge of the field to complete the sublimation of the gardening spirit. It is like a three-dimensional record of China's agricultural civilization and horticultural life, adding radiance and beauty together with the world garden, silently watching our home.

 Inner Nature Imagery

轻落
溪水湖畔的演艺中心

觅——当蝴蝶轻轻落下，这里就是她寻找的美丽家园。承担着本次世园会开、闭幕式以及重要国家日等演出活动的演艺中心，如一只展翅欲飞的彩蝶，被园区春色绵绵、绿意盎然的湖景画卷吸引，停驻在天田山东侧妫汭湖边，遥看永宁阁，相望中国馆，尽享湖景山色。

轻——结合建筑的彩蝶形象，周边景观采取“轻设计”的理念。从使用材料、绿化种植形式等多方面凸出建筑主体，融入周边环境，切合办园主题，营造绿色、质朴、放松的景观效果。舞台、看台及坐凳、二层平台等必要硬质功能区多使用防腐木、石材碎拼、露骨料混凝土等作为面层材料，体现出材料之“轻”；绿地则以大面积绿坡草坪、点植遮荫大树、镶边观花乔灌作为种植策略体现空间之“轻”，在结合彩色遮阳屋顶的半开放空间中营造了座席区、集散休憩活动平台及步道，实现贴近自然、听风看雨、闻香赏花的游览及观演体验。

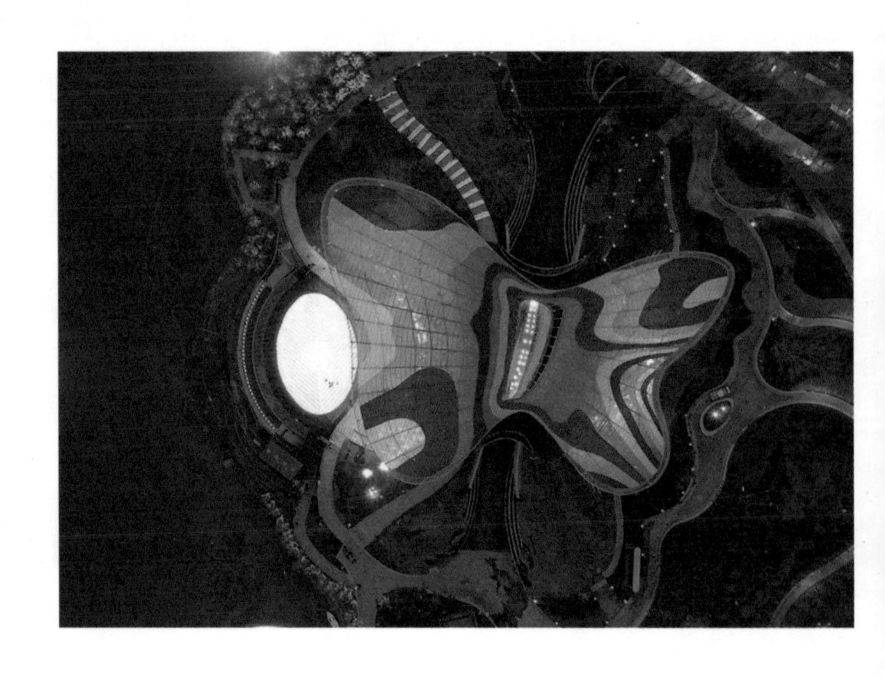

Put Down Slightly — Performing Arts Center Beside Lake

Finding — when the butterfly drops gently, here is the beautiful home she seeks. The performing arts center, which is responsible for the opening and closing ceremonies of the world garden expo as well as the performance activities for important national days and other events, is as a colorful butterfly that is going to fly. It is attracted by the picture scroll of lake scenery full of spring scenery and green in the garden area, stopping and standing near the Guirui Lake in the east of Tiantian Mountain, looking at the Yongning Pavilion in the distance, looking at each other with China Pavilion, and enjoying the lake scenery.

Light — combined with the colorful butterfly image of the building, the surrounding landscape adopts the concept of "light design". Highlight the main building from used materials, green planting forms and other aspects, blend in with surroundings, conform to the theme of the garden, and create green, plain, relaxed landscape effect. The necessary hard functional areas such as the stage, the stand, the bench and the second-floor platform usually use antiseptic wood, broken stone, and exposed concrete, etc. as surface material to reflect the "lightness" of the material; the green land usually use large area of green slope lawn, planting separately shade trees, planting flowers, arbors and shrubs as planting strategy to reflect the "lightness" of the space. Combined with the semi-open space of the colorful sunshade roof, the seating area, gathering and distributing activity platform and walkway are created to realize the sightseeing and performance experience close to nature, listening to the wind, watching the rain, smelling the fragrance and appreciating the flowers.

融入
天然的生活场景

世园会生活体验馆，位于国际展区北部原始林地保护地，毗邻刺槐林，一条原生旱柳林道贯穿场地。呼应2019年世界园艺博览会“园艺融于自然，自然感动心灵”的办会理念，生活体验馆借助周边原始环境的优势，在营建中秉持“自然、融入、人性，体现场所精神”的设计原则，通过嵌入、随行、顺色的造园工法对场地形态、竖向地形、建筑群落、街巷漫道、草坡林荫等空间进行"场所"适地设计。

场地因借旱柳林道南北走向强化出的横纵8条道路分割出16个方格矩阵，16个方格由东北角4个方格就近结合形成一号馆，向西南角等距展开形成二号馆到五号馆，共5个场馆空间，再向西南方向展开营建有机田园风貌。因此，这片沃土因为与旁边槐树树林的连绵而没了边界的束缚，在阡陌交错中有的是千米麦田、有的是百米海棠、有的是九月粉色浪舞的观赏草海，利用道路划分出的地块田头以园艺色彩拼绘出季节的绚烂。

Integrate — Scenes of Daily Life in natural

The World Horticulture Exposition Life Experience Hall is located in the original woodland protection area in the north of the International Exhibition Area, adjacent to the Robinia forest and a road of native Salix Matsudana forest throughout the site. Echoing the holding concept of the 2019 World Horticulture Exposition of "Gardening in nature, nature touches the heart", the Life Experience Pavilion uses the advantages of the surrounding primitive environment to build a design principle upholds nature, integration, humanity and reflects the spirit of the place. The site form, vertical topography, building community, streets and lanes, grass slopes, and tree shade are used to design "places" on the spot through the embedded, accompanying, and smooth gardening method.

Due to the South-North trend of the Salix forest road, the site is divided into eight vertical and horizontal roads, which are divided into 16 square matrix. The 16 square grid comprises four squares in the northeast corner to form hall 1. They Expand equidistantly to the southwest corner to create No. 2 to No. 5, a total of five venues, and then to the southwest direction of the idyllic organic scenery. Therefore, when the fertile land connects with the nearby Acacia forest, the boundary disappears; thousands of Wheatland, hundreds of meters of Begonia, and the pink grass sea fluctuate with the wind in September so that the fields divided by roads are painted with gardening color to show the splendor of the season.

Inner Nature Imagery

心象自然后记

2002年底，一直从事规划和建筑设计的我和史丽秀，向院里提出组建景观设计所，以此适应当时方兴未艾的园林景观发展趋势。

景观所创立之初，一切从零开始，项目少、人员不足、收入也不高，发展得很艰难，但是我们始终有一个信念，就是要坚持思考，用研究的态度对待每一个项目，坚持做研究性设计。这个信念，帮助我们抵挡了很多单纯赚钱的项目，也为我们承接的每个项目增加了不少工作量，但我们乐此不疲。在这种坚持下，慢慢形成了景观所的一种基因，一种文化，大家都有所追求地对待每一个项目。正是这种坚持，使我们逐渐赢得了口碑，赢得了业主，赢得了越来越好、越来越大的项目。

在研究性设计中，我慢慢地从对项目的思考转向对园林景观设计的思考，希望设计能够回归其本源，回归到适应性的人本需要、创新性的艺术追求、生态性的自然关怀。再后来，我意识到园林景观设计的本质是平衡人和自然的关系。在设计中，要解决的关键问题是平衡主观和客观的关系。于是，“心象自然”的设计观渐渐浮现出来。也许是研究生期间对主体论和方法论的偏爱，我更关心设计师如何去设计，用什么观点、什么态度、什么方法、什么追求去设计。“心象自然”使我能比较恰当地描述人和自然的关系、主观和客观的关系，从而能为景观设计方法论找寻依据。

本书的写作不是简单的项目介绍，而是按照“心象”、“自然”和“心象自然”三个篇章，从“心象篇”理性和感性的外化、“自然篇”生态和文明的表现、“心象自然篇”主观与客观的平衡以及人与自然的和谐等逐层递进展开阐述。所列举的项目实例是从大量建成作品中挑选出来的，一方面为了配合文字内容体现“心象自然”的理念和方法，一方面也能客观地反映出景观所创立近二十年来的研究性设计成果。

本书成稿过程中参与部分文字写作的有关午军、赵文斌、刘环、路璐、朱燕辉、杨陈、巩磊、雷洪强、贾瀛等，尤其是关午军为文字和图片的整理编排付出了大量的辛苦，功不可没。感谢张广源、王祥东等提供的大量照片；感谢编辑付娇、兰丽婷等的辛勤付出；感谢曹静宜、刘宇婷、韩迅的翻译工作。

感谢多年来关心和支持我们的各级领导、院里的各位老总、每个项目的业主以及共同完成这么多项目的所有景观所的同事和项目合作伙伴，是你们成就并见证了景观所的成长与发展，也见证了我们不断思考不断研究的心路历程。

李存东

Afterword of Inner Nature Imagery

At the end of 2002, Shi Lixiu and I, who had been engaged in planning and architectural design, proposed to set up a landscape design institute to adapt to the development trend of garden landscape which is in the ascendant at that time.

When the landscape design institute was created at the beginning, it was difficult to start from scratch, with few projects, insufficient staff and low income. But we always have a belief that we should keep thinking, treat every project with the attitude of research, and insist on doing research design. This belief has helped us fend off a lot of purely profitable projects and has added a lot of work to every project we undertake, but we always enjoy it. In this kind of persistence, it gradually formed a kind of gene of the landscape design institute, a kind of culture, that everyone has something to pursue to treat every project. It is this kind of persistence that makes us gradually earn the reputation, earn the owners, earn the better and better, bigger and bigger projects.

In the research design, I gradually changed from thinking about the project to thinking about the garden landscape design, hoping that the design could return to its original source, returning to the adaptive humanistic needs, innovative artistic pursuit and ecological natural care. Later, I realized that the essence of garden landscape design is to balance the relationship between man and nature. In the design, the key problem to solve is to balance the relationship between subjective and objective. Thus, the design concept of "inner nature imagery" gradually emerged. Perhaps due to my preference for subjectivism and methodology during my postgraduate study, I am more concerned about how designers design, with what kinds of views, attitudes, methods and pursuits."Inner nature imagery" enables me to describe aptly the relationship between man and nature, the relationship between subjectivity and objectivity, so as to find the basis for the methodology of landscape design.

This book is not a simple project introduction, but according to three chapters of "Inner Imagery", "Nature Imagery" and "Inner Nature Imagery" , from externalization of reason and emotion in the "Inner imagery" ,expression of ecology and civilization in the "Nature Imagery", balance between subjectivity and objectivity in the "Inner Nature Imagery" and harmony between human and nature and so on, expand the exposition step by step. The project examples listed are selected from a large number of completed works. On the one hand, they are designed to match the text content to reflect the concept and method of "inner nature imagery", and on the other hand, they can objectively reflect the research design results of nearly two decades since the establishment of the landscape design institute.

In the process of finishing this book, Guan Wujun, Zhao Wenbin, Liu Huan, Lu Lu, Zhu Yanhui, Yang Chen, Gong Lei, Lei Hongqiang, Jia Ying and so on participated in part of words writing, in particular, Guan Wujun paid a lot of hard work for the arrangement of text and picture, whose contributions cannot go unnoticed.Special appreciation for the massive photos providing by Zhang Guangyuan, Wang Xiangdong, etc. editing work by Fu Jiao, Lan Liting, etc. and also translation job from Cao Jingyi, Liu Yuting, and Han Xun.

Many thanks for every leaders at all levels, every chief designers in the landscape design institute, owners of every project, who care for us and support us for many years, as well as colleagues in the landscape design institute and project partners who worked together on so many projects. It is you who have accomplished and witnessed the growth and development of the landscape design institute, as well as the process of our continuous thinking and research.

Li Cundong

图书在版编目（CIP）数据

心象自然 = Inner Nature Imagery：汉英对照 / 李存东，史丽秀著. — 北京：中国建筑工业出版社，2020.10

ISBN 978-7-112-25288-6

Ⅰ.①心… Ⅱ.①李…②史… Ⅲ.①景观设计－作品集－中国－现代 Ⅳ.①TU983

中国版本图书馆CIP数据核字（2020）第114910号

责任编辑：兰丽婷
书籍设计：张悟静
责任校对：王 烨

心象自然
Inner Nature Imagery
李存东 史丽秀 著
*
中国建筑工业出版社出版、发行（北京海淀三里河路9号）
各地新华书店、建筑书店经销
北京富诚彩色印刷有限公司印刷
*
开本：787毫米×1092毫米 1/16 印张：20 插页：5 字数：485千字
2021年4月第一版 2021年4月第一次印刷
定价：225.00 元
ISBN 978-7-112-25288-6
（36082）